ELEMENTARY KINEMATICS OF MECHANISMS

ELEMENTARY KINEMATICS OF MECHANISMS

by
JOHN R. ZIMMERMAN
Assistant Professor of Mechanical Engineering
The Pennsylvania State University

JOHN WILEY & SONS, INC.
NEW YORK · LONDON

Library of Congress Catalog Card Number: 62-8767

Printed in the United States of America

Preface

This textbook is designed for the student of limited mathematical background who is making his first study of the analysis and design of mechanisms. It is presupposed that the student brings with him competence in the elements of algebra and trigonometry and a knowledge of basic drafting technique. Although it would be well if he has had a course in rigid body mechanics, the book contains sufficient introductory material on vectors and the more elementary kinematic relationships to make its use quite possible for those lacking a background in mechanics. No knowledge of calculus or advanced mathematics is required.

Emphasis is on the fundamentals of motion analysis and mechanism design rather than on descriptive detail. This does not imply that I regard the descriptive detail on such important elements as gears, cams, and flexible connectors as unimportant; on the contrary, the student should have access to and be urged to consult the well-written and lavishly illustrated publications of our industrial firms. He must develop an acquaintance both with basic principles of analysis and design and with the "hardware."

Graphical techniques have been used very often, for they are usually the more economical of effort and time. However, no deliberate attempt has been made to avoid the use of simple algebraic methods when they seem to be justified. The modern designer can ill afford strong prejudices on this matter of the graphical "versus" the algebraic approach. Whichever method achieves a sufficiently accurate solution with the least labor and mental torment is the one to use. Indeed, often it is a blend of the graphical and the algebraic approaches that is most fruitful.

Although an ample number of problems has been given in the text, one of the many excellent problem sets for kinematics now available can be used to advantage both for homework and for practicum assignments.

I shall greatly appreciate hearing from the users of this book concerning both errors and suggestions for improvements.

My debts are greatest to a number of men known to me only through their writings. To Rudolf Beyer and Kurt Hain of Germany, to N. Rosenauer and A. H. Willis of Australia, and to the Americans, Joseph S. Beggs, Ferdinand Freudenstein, A. S. Hall, Jr., A. E. Richard de Jonge, Harold A. Rothbart, and Joseph E. Shigley, grateful acknowledgment is made.

And, finally, to my wife Grace, and to my children Jan and Scott, an affectionate acknowledgment is made for putting up with it all so cheerfully.

JOHN R. ZIMMERMAN

University Park, Pennsylvania
December, 1961

Contents

Symbols and Abbreviations

Symbol	Meaning
$\mathbf{a}$ or $\mathbf{A}$	acceleration
a or A	usually magnitude of acceleration; for rectilinear motion it designates acceleration
$\mathbf{a}^n$ or $\mathbf{A}^n$	normal component of acceleration
$\mathbf{a}^t$ or $\mathbf{A}^t$	tangential component of acceleration
$\mathbf{a}^c$ or $\mathbf{A}^c$	Coriolis component of acceleration
a_{12}	midnormal of line segment A_1A_2; b_{12} is midnormal of segment B_1B_2, etc.
a	addendum
b	dedendum
B	backlash
C	center distance
d	diameter
D	diameter; pitch diameter
f	proper fraction
F_i	input force
F_o	output force
h_t	whole depth of gear tooth
h_w	working depth of gear tooth
k_a	acceleration scale
k_s	displacement scale
k_t	time scale
k_v	velocity scale
L	connecting rod length
L	belt length
m_g	gear ratio
m_p	contact ratio
n	number of threads on worm
n	number of links
N	number of instant centers
N	number of teeth
N	angular velocity in revolutions per minute
O_{12}, O_{57}, etc.	instant centers
p	circular pitch
p_b	base pitch
p_n	normal circular pitch
P	diametral pitch
P_n	normal diametral pitch
P_{12}, P_{23}, etc.	virtual poles
P_i	input power
P_o	output power
r	radius
R	radius; pitch radius of gear
R_b	base radius of gear
R_{12}, R_{23}, etc.	relative poles
Q	length of arc of action
$\mathbf{s}$	linear displacement
s	distance; magnitude of linear displacement; for rectilinear motion it designates displacement
s	percent slip
t	time
t	circular tooth thickness
t	belt thickness
T	period
T_i	input torque
T_o	output torque
$\mathbf{v}$ or $\mathbf{V}$	velocity
v or V	usually speed; for rectilinear motion it indicates velocity
Z	length of path of action
α (alpha)	pitch angle of bevel pinion
α (alpha)	angular acceleration (positive if counterclockwise)

α (alpha)	pressure angle in cam mechanism
β (beta)	pitch angle in bevel gear
Δ (delta)	when preceding a quantity indicates a change in that quantity
θ (theta)	angular position; angular displacement
ϕ (phi)	angular position; angular displacement
ϕ (phi)	pressure angle of gear
ϕ_N	normal pressure angle of a gear
ψ (psi)	angular position; angular displacement
ψ (psi)	helix angle
ω (omega)	angular velocity (positive if counterclockwise)
ω_i	angular velocity of input link
ω_o	angular velocity of output link

BDC	bottom dead center
CA	constant acceleration
CR	connecting rod
ccw	counterclockwise
cw	clockwise
D-R-R-D	dwell, rise, return, dwell
D-R-D-F-D	dwell, rise, dwell, fall, dwell
log	logarithm to the base 10
LH	left hand
MCV	modified constant velocity
RH	right hand
MA	mechanical advantage
mph	miles per hour
R-R-R	rise, return, rise
rev	revolution
rpm	revolutions per minute
rps	revolutions per second
SHM	simple harmonic motion
TDC	top dead center

ELEMENTARY KINEMATICS OF MECHANISMS

I Kinematic Terminology

1-1. Introductory

The art of mechanical design is directed toward the creation of new or improved machines and instruments to modify and control energy and motion. The designer accepts the task of planning and designing equipment that will not only perform the functions expected of it but will also do so with the greatest possible over-all economy. The designer's responsibility does not cease with the conception of a machine which will perform the functions expected of it; he must consider as many aspects of the machine's manufacture and use as possible, problems of ease of manufacture, appearance, safety, and so on.

Design is, therefore, a complicated enterprise; it necessitates a considerable exercise of judgment and a certain amount of trial and error. Design can never be reduced to a mechanical routine. Yet there is a pattern to it. Figure 1-1 illustrates this pattern. Although the diagram shows the logical sequence of steps in design, it must not be supposed that they always proceed in precisely this order or that they proceed independently of one another. Changes in one aspect of the design usually entail changes in others.

Even when the machine has gone into final production, the designer is not yet quite through, for there is much to be learned by keeping in touch with those responsible for its manufacture and with the machine's performance in actual service. This information can be of incalculable benefit in future work. One of the hallmarks of intelligence is the ability to learn from experience, particularly from one's mistakes.

1-2. Kinematics of Mechanisms

Kinematics of mechanisms is that discipline which deals with the motions of mechanisms, without regard to the forces which affect these motions. Two aspects of the subject can be distinguished. An investigation of the motion characteristics of a mechanism of existing design is *kinematic*

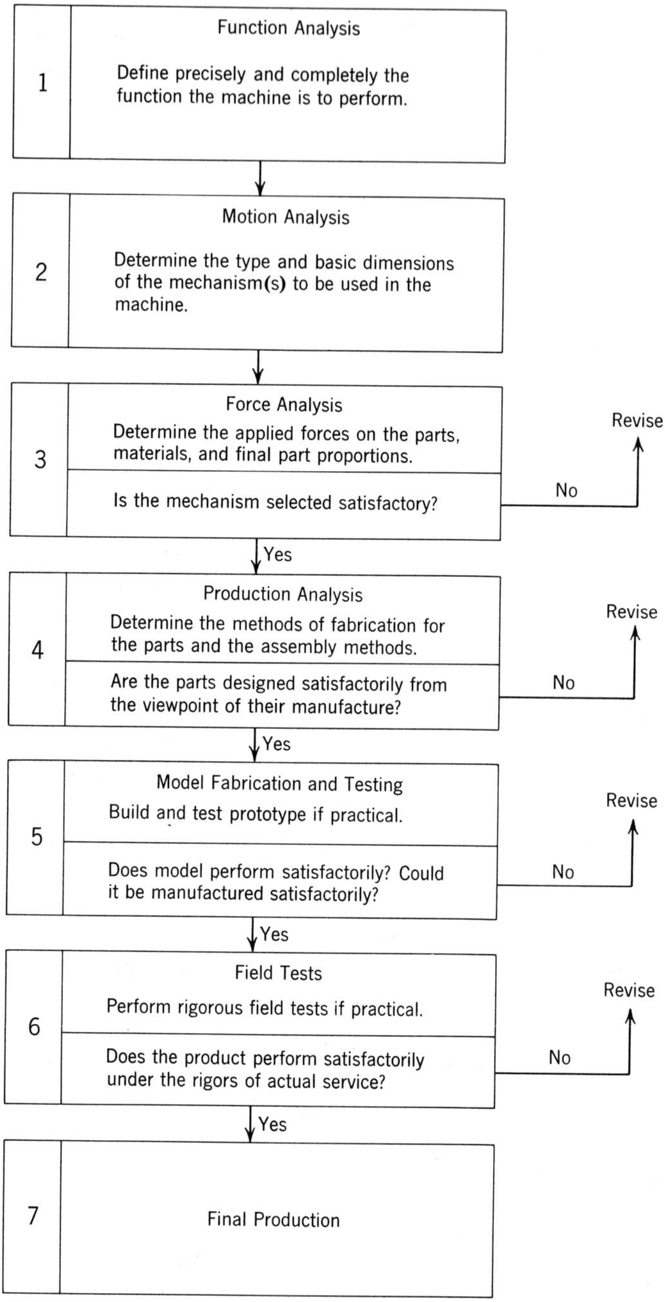

Fig. 1-1. Mechanical design procedure.

analysis. On the other hand, the design of a mechanism to provide certain desired motions is *kinematic synthesis*.

1-3. Mechanisms and Machines

A *mechanism* is an assembly of bodies, usually rigid, such that the motion of one of the bodies causes definite and predictable motions of the other bodies in the assembly. A *machine* is a mechanism or combination of mechanisms capable of transmitting and modifying mechanical energy. The primary purpose of the mechanism is to transmit and modify motion, while the machine has the additional function of transmitting and modifying mechanical energy. The internal combustion engine, for example, converts the work energy given to the pistons by the burning gases into work done by the rotating shaft driving an external load. A watch mechanism, on the other hand, is not designed to transmit energy. The proper movement of its hands is its sole function. The energy stored in the watch spring is used only to overcome the frictional resistances encountered by the members of the mechanism.

1-4. Links and Pairs

A *link* is a member of a mechanism connecting other links and having motion relative to them. Links may be supports, may guide the motion of other links, transmit motion, or serve simultaneously in more than one of these capacities. The simplest mechanisms have three links: driver, driven, and frame.

Figure 1-2 illustrates a simple three-link cam mechanism. Link 1 is the frame and supports links 2 and 3. Link 2 is the driver and imparts motion to the driven link 3.

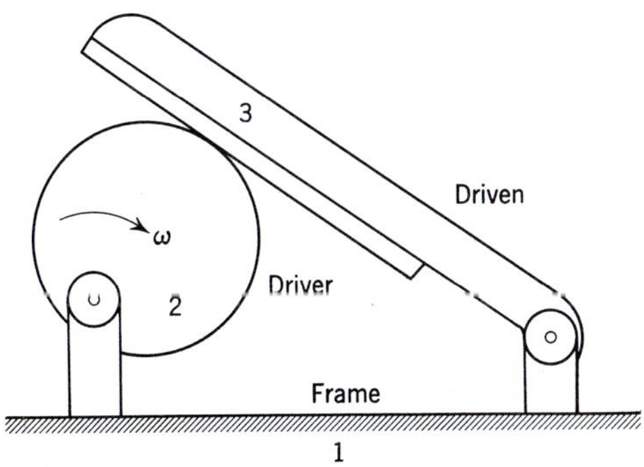

Fig. 1-2. Three-link cam mechanism.

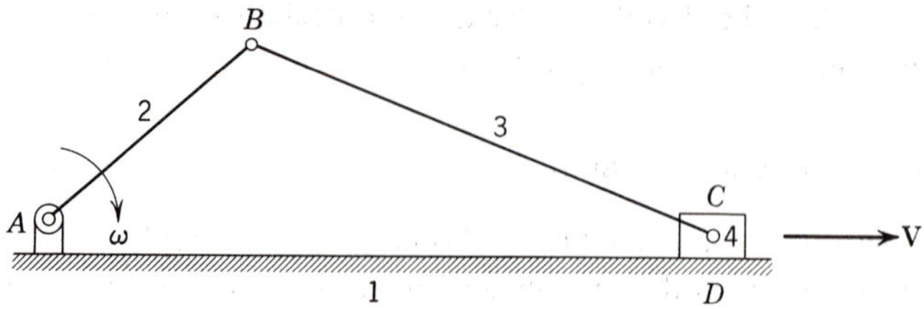

Fig. 1-3. Slider-crank mechanism.

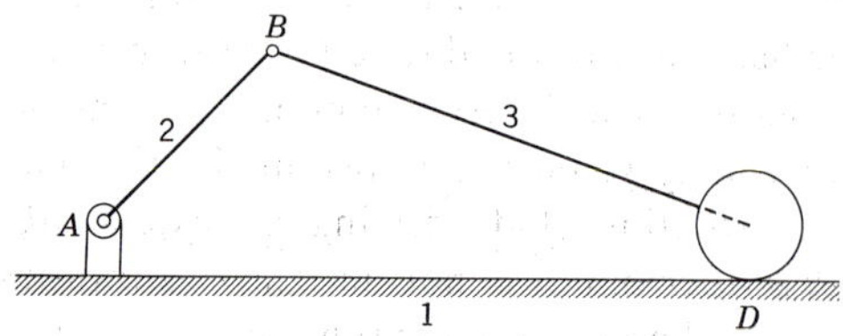

Fig. 1-4. Slider crank with ball slider.

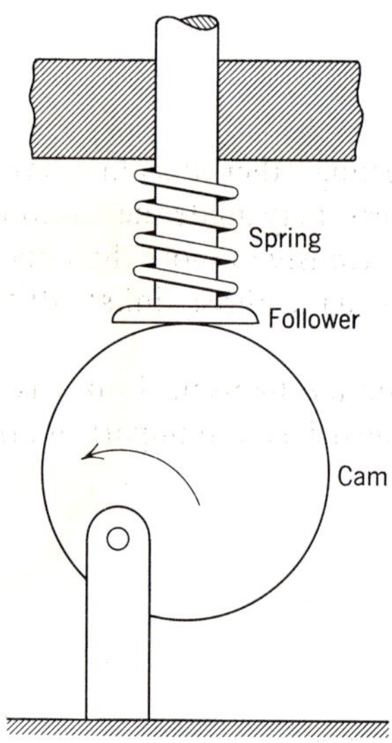

Fig. 1-5. Cam mechanism illustrating unclosed pair.

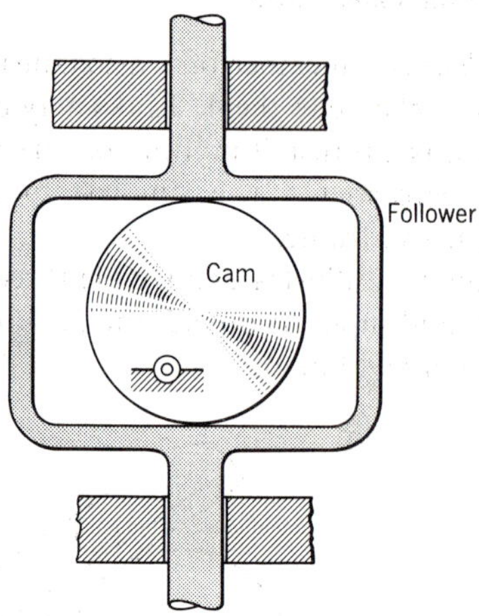

Fig. 1-6. Cam mechanism illustrating closed pair.

In Fig. 1-3 one of the most useful four-link mechanisms is shown, the familiar slider-crank mechanism used in reciprocating engines and compressors. Link 1 is the fixed member; it supports links 2 and 4 as well as serving to guide the motion of link 4, the slider. Link 2 is, if this is a compressor, the driver; it transmits motion to the connecting rod, link 3, which in turn transmits motion to the driven member, the slider, link 4.

A *pair* is a joint that permits relative motion. A pair is made up of two elements, one on each link being joined. In Fig. 1-3, for example, there is a turning pair at A: a joint permitting relative rotation of the crank and connecting rod. At D, however, there is a sliding pair: a joint permitting relative sliding between the slider and the guide. In a roller bearing there is rolling contact between the roller and each of the two races.

Pairs are classified as higher or lower. A *higher pair* is one in which there is point or line contact between the links being joined. A wheel rolling on a surface, for example, is an instance of a higher pair. A *lower pair* is one in which there is surface contact between the members being joined. A shaft rotating in a bearing, or a nut turning on a screw, is an example of lower pairing.

Frequently higher and lower pairs can be interchanged without affecting the relative motions of the links. In Fig. 1-4 the slider-crank mechanism has been redesigned by replacing the lower pairs at D by a higher pair. A spherical surface on the connecting rod has been substituted for the piston (thus eliminating one link), with the result that point contact between the spherical surface and the guide replaces the surface contact between the slider and guide. The design in Fig. 1-4 will be inferior from the standpoint of wear to that of Fig. 1-3, but from the viewpoint of kinematic behavior, there is no difference.

Pairs are also classified as closed and unclosed. A *closed pair* is one in which the links are held together mechanically, where only the destruction of one or more members could make possible a breaking of contact. An *unclosed pair* is one that relies upon some external means, such as a spring or gravity, to maintain contact. Figures 1-5 and 1-6 illustrate the difference. In Fig. 1-5 the follower is maintained in contact with the cam by the spring, while in Fig. 1-6 the follower is designed in such a way that contact is assured.

1-5. Kinematic Chains and Inversion

When a group of links are joined together so that, if one of the links were fixed, the others would move in definite and predictable ways, the assembly is called a *kinematic chain*. A mechanism is made from a kinematic chain by fixing one of the links; therefore there are as many possible mechanisms as there are links. The various mechanisms that can be made from a kinematic chain are its *inversions*.

This can be illustrated by noting the possible inversions of the sliding-block chain shown in Fig. 1-7. Table 1-1 lists each of the four possible inversions of this chain with common applications.

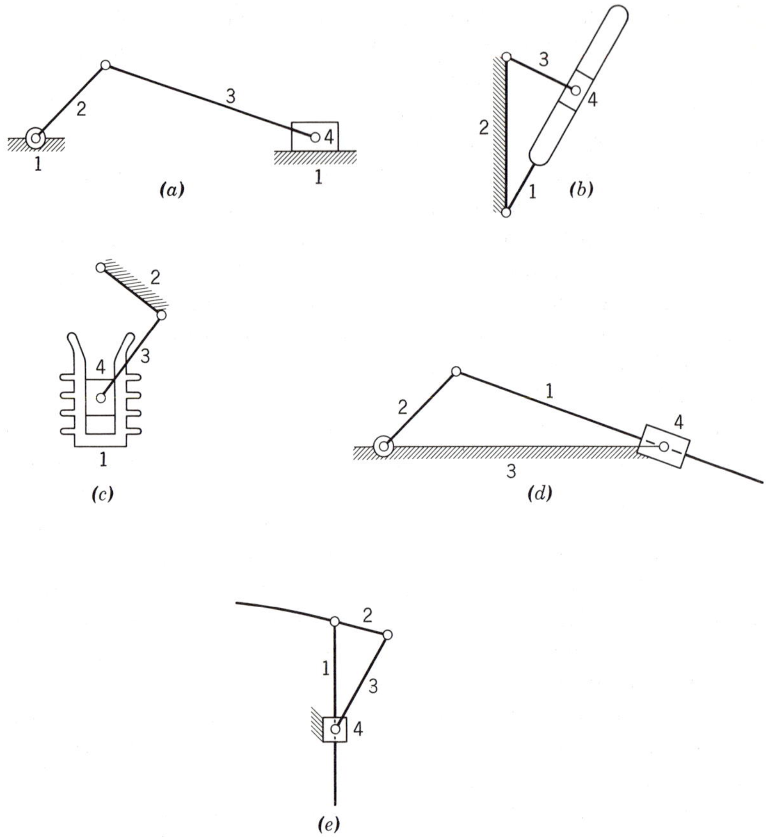

Fig. 1-7. Inversions of the sliding block chain.

TABLE 1-1

Inversions of the Sliding-Block Chain

Inversion	Figure	Fixed Link	Applications
1st	1-7*a*	1	Reciprocating engines and compressors
2nd	1-7*b*	2	Quick-return
	1-7*c*	2	Rotary engine
3rd	1-7*d*	3	Small oscillating steam engine
4th	1-7*e*	4	Hand pump (not a widely used inversion)

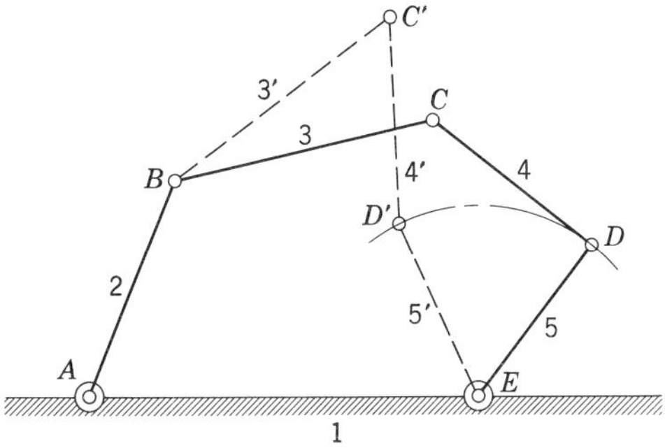

Fig. 1-8. Unconstrained chain.

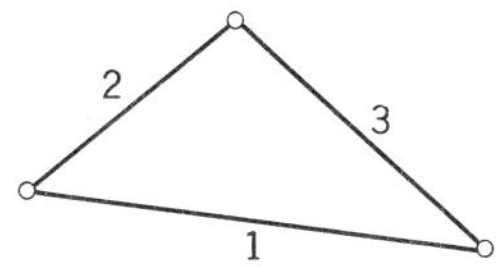

Fig. 1-9. Structure.

The definition of kinematic chain included the qualification that, with one link fixed, the others are obliged to move in definite ways. Unless this condition is adhered to, the assembly will fail to be a kinematic chain. An example of this is shown in Fig. 1-8. Here is a five-link assembly which fails to qualify. To test it, one of the links (link 1) has been fixed. For a single position of link 2, there are an indefinite number of positions in which links 3, 4, and 5 might be arranged; two of which have been shown. This chain is *unconstrained*. To give it constrained motion, a second link, such as link 5, could be used as a second input link.

The assembly of Fig. 1-9 fails also, but for another reason. There can be

Fig. 1-10. Jaw crusher mechanism. (Courtesy Allis-Chalmers Manufacturing Co., Milwaukee, Wisc.)

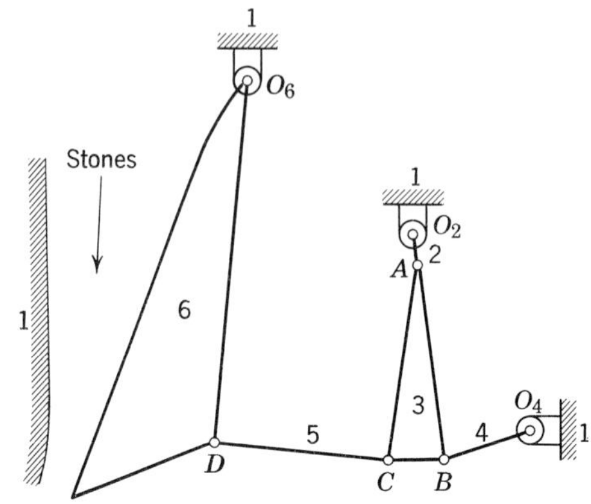

Fig. 1-11. Kinematic representation of jaw crusher mechanism.

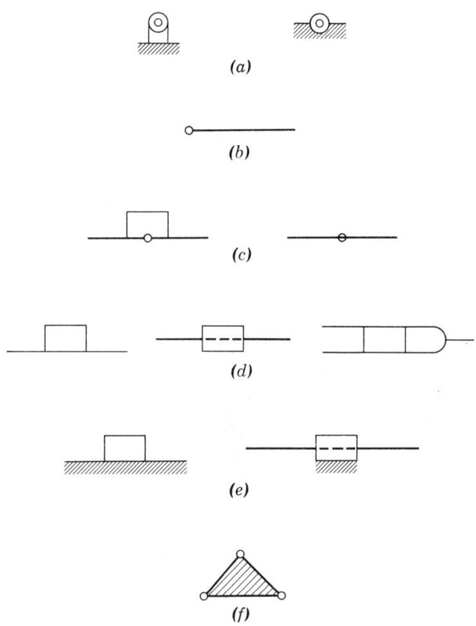

Fig. 1-12. Some kinematic symbols. (*a*) Fixed axis of rotation. (*b*) Pin joint at end of link. (*c*) Pin joint on link. (*d*) Sliding pairs. (*e*) Fixed guide. (*f*) Rigid link with three noncollinear pin joints.

no relative motion of the members. It is a structure; and, as a structure, it could serve as a single link in a kinematic chain, but not as a kinematic chain.

1-6. Schematic Diagrams

To take the trouble to show the members of a mechanism in their true shapes would be a laborious and unnecessary chore. Only a few of the dimensions of the members of a mechanism are of any relevance to a kinematic analysis. For example, in Fig. 1-10 a jaw crusher mechanism is shown in great detail. The kinematic representation of this mechanism in Fig. 1-11 is a sharp contrast in its stark simplicity. An enormous amount of detail has been omitted in making this kinematic representation, yet it shows all the dimensions essential to a motion analysis of the mechanism.

A number of the most commonly used kinematic symbols are shown in Fig. 1-12.

REFERENCE

1. R. S. Hartenberg and J. Denavit, "Men and Machines," *Machine Design*, May 3, June 14, and July 12, 1956.

2 Basic Principles of Motion Analysis

2-1. Introduction

It is the chief purpose of this chapter to review the basic principles of general kinematics necessary for future work in the kinematics of mechanisms. Vectors, their addition, subtraction, and resolution, the simpler types of motion, and a few topics of special interest will be covered. For the student who lacks a background in mechanics, the topics dealt with here are designed to give the necessary foundation for an understanding and appreciation of what is to come later. For the student with prior training in mechanics, much of this material will be a review.

2-2. Vectors

Scalar and vector quantities. Those physical quantities which can be represented simply by numbers are referred to as *scalar quantities*. Examples of scalar quantities are mass, temperature, length, and electric charge. Physical quantities which involve the notion of direction as well as magnitude are called *vector quantities*. Quantities such as force, acceleration, and moment are examples of quantities where the specification of a direction is essential.

In Fig 2-1 the flight of a plane is illustrated. The plane started at point A and flew 200 miles to a point B. The displacement of B relative to A is 200 miles N 30° E, a vector quantity. If the direction of the plane's flight were not mentioned, it would only be known that B is somewhere on a circle of 200 miles radius with center at A. Specifying both the direction and the distance of B from A is necessary if its exact location is to be given.

Two vector quantities are equal if, and only if, they have the same magnitude and the same direction. The vector quantity $k\mathbf{A}$, where k is a scalar, is k times as great as $\mathbf{A}$ in magnitude, but in the same direction if k is positive, or opposite if k is negative.

Representation of a vector quantity. A vector quantity is represented graphically by a directed line segment called a *vector*. The length of the vector represents to some arbitrary scale the magnitude of the vector quantity, and its direction is that of the vector quantity.

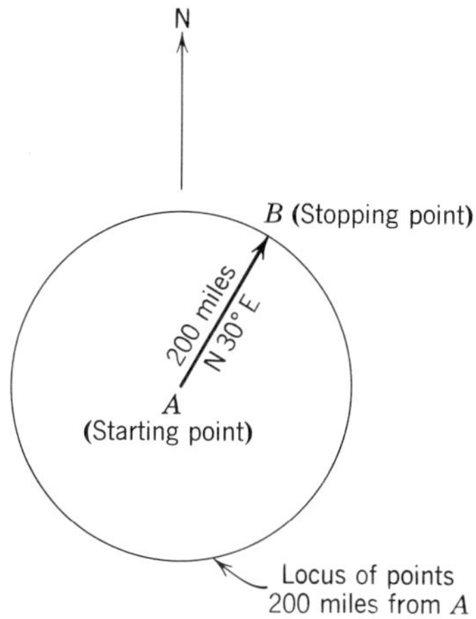

Fig. 2-1. Displacement is a vector quantity.

A vector **A** is represented in Fig. 2-2. In Fig. 2-2*a* x- and y-axes have been drawn through the origin of the vector. Its angular position has been specified by the angle the vector makes with the x-axis. A simpler way is shown in Fig. 2-2*b*. It is understood that the angle indicated is the angle made by the vector with an x-axis through its origin.

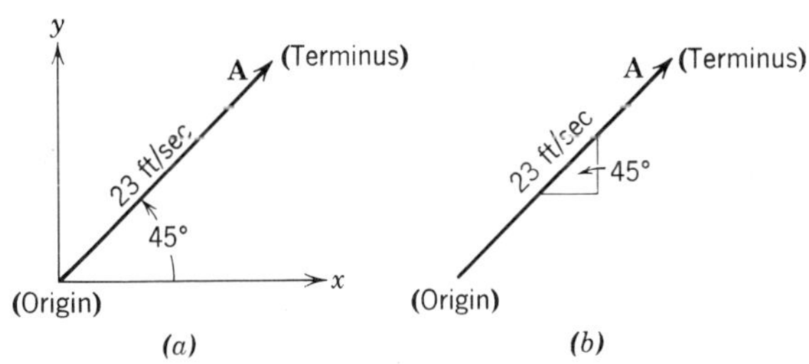

Fig. 2-2. Representation of a vector quantity.

Vector addition. The addition of vectors is illustrated in Fig. 2-3. Vectors **A** and **B** are given. It is required to find **C**, where

$$\mathbf{C} = \mathbf{A} + \mathbf{B}$$

The triangle or the parallelogram methods can be used.

Triangle Method. Draw one of the vectors. Draw the second with its origin at the terminus of the first. The resultant (vector sum) is the vector from the origin of the first to the terminus of the second. This is illustrated in Fig. 2-3*b*.

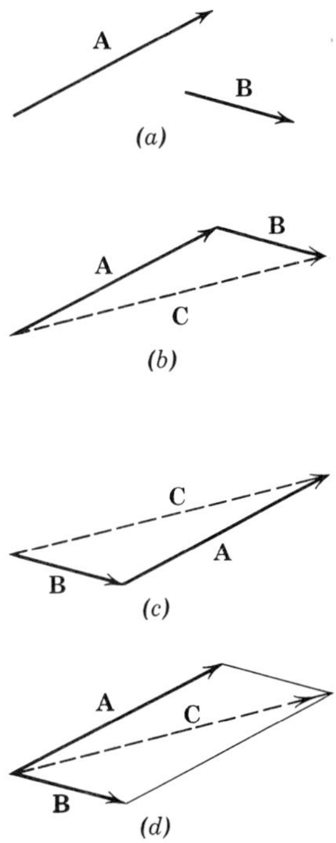

Fig. 2-3. Addition of two vectors: **C** = **A** + **B**.

The order in which the vectors are added has no effect on the result: that is, vector addition is *commutative*. In Fig. 2-3*c* the vectors have been added, using the triangle method, but in a different order. The result is the same.

Parallelogram Method. Draw the vectors so that their origins coincide, as in Fig. 2-3*d*. From the termini of the two vectors, draw lines to complete the parallelogram. The resultant is the diagonal of the parallelogram from the common origin. The sense of the resultant is away from the common origin.

When more than two vectors are to be added, an extension of the triangle method, the polygon method, is used. Figure 2-4*a* shows four vectors, **A**, **B**, **C**, and **D**. It is required to find **R**, where

$$\mathbf{R} = \mathbf{A} + \mathbf{B} + \mathbf{C} + \mathbf{D}$$

The solution is shown in Fig. 2-4*b*.

> *Polygon Method.* Draw any of the vectors. Draw a second vector with its origin at the terminus of the first. Continue, drawing each vector with its origin at the terminus of the one previously drawn, until all have been drawn. The resultant is the vector from the first origin to the last terminus.

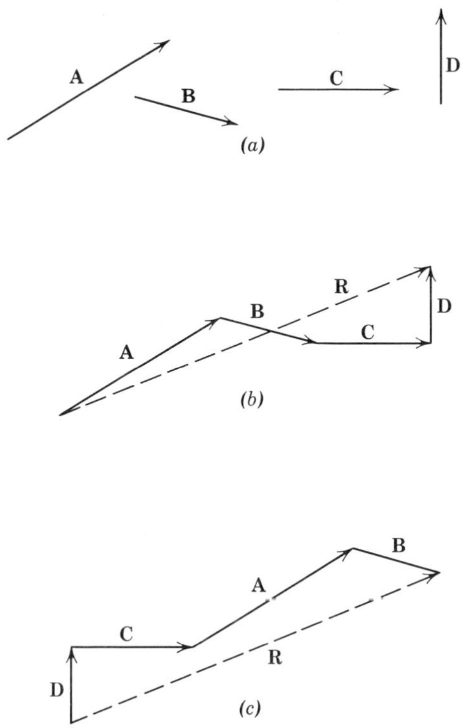

Fig. 2-4. Addition of more than two vectors: $\mathbf{R} = \mathbf{A} + \mathbf{B} + \mathbf{C} + \mathbf{D}$.

The order in which the vectors are drawn makes no difference to the final result. In Fig. 2-4*c* the vectors have been added in a different order from that used in Fig. 2-4*b*. However, the result is the same.

Vector subtraction. When motion analyses of mechanisms are made, there is frequently occasion to perform a vector subtraction. Three methods are available. Each is the most convenient on some specific occasion; consequently each should be learned. In Fig. 2-5*a* there are two vectors, **A** and **B**. It is required to find **C**, where

$$\mathbf{C} = \mathbf{A} - \mathbf{B}$$

Method 1. Draw the two vectors so that their origins coincide. Their vector difference, **C**, is the vector from the terminus of **B** toward the terminus of **A**. This solution is shown in Fig. 2-5*b*.

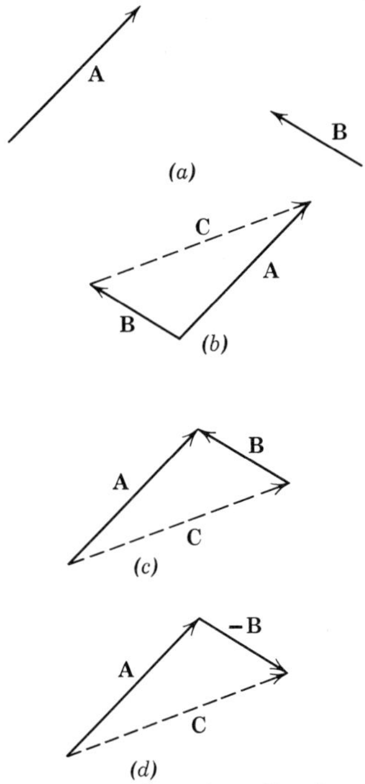

Fig. 2-5. Vector subtraction: **C** = **A** − **B**.

Method 2. Draw the two vectors so that their termini coincide. Their vector difference, **C**, is from the origin of **A** toward the origin of **B**. The solution is shown in Fig. 2-5*c*.

Method 3. Add **A** and −**B**. This is the easiest of the three to remember, but not always the most convenient to apply. The solution is shown in Fig. 2-5*d*.

Resolution of a vector. In the methods of vector addition discussed previously, a number of vectors were given; and it was desired to find a single vector, their resultant, to replace them. Resolution of a vector has the opposite purpose, the replacing of a single vector by a group of vectors. Usually it is a pair of vectors which is required. However, since there are

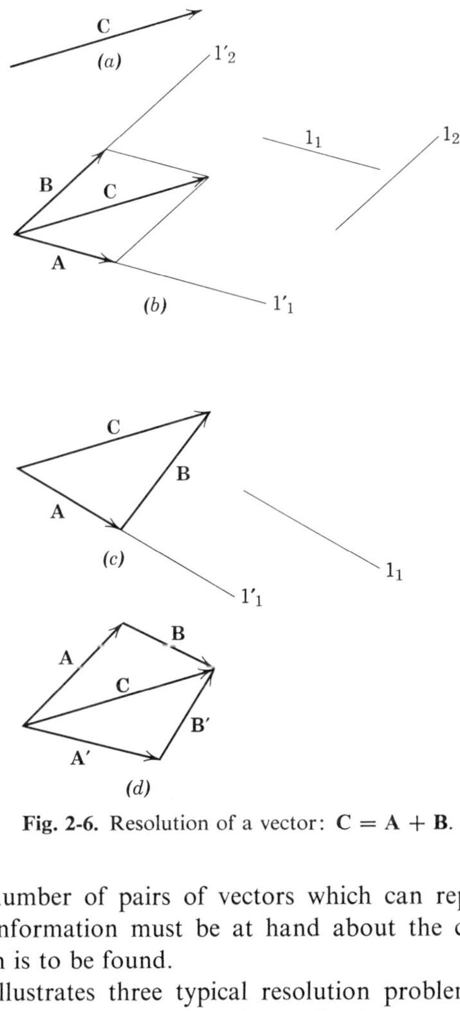

Fig. 2-6. Resolution of a vector: $\mathbf{C} = \mathbf{A} + \mathbf{B}$.

an indefinite number of pairs of vectors which can replace any single vector, some information must be at hand about the components if a unique solution is to be found.

Figure 2-6 illustrates three typical resolution problems that arise in practice. Vector **C** in Fig. 2-6*a* is to be resolved.

In Fig. 2-6*b* the directions of the two components are known. One is to be parallel to line 1_1 and one to 1_2. Lines $1'_1$ and $1'_2$ are drawn through the origin of **C** and parallel to 1_1 and 1_2. Then lines are drawn from the terminus of **C** and parallel to l_1 and l_2 to complete the parallelogram. Components **A** and **B** are thus determined.

In Fig. 2-6*c* the magnitude of component **A** is known, and it is known to have the direction of line l_1. From the origin of **C**, vector **A** is laid out with the given direction and magnitude. Then component **B** is the vector from the terminus of **A** to the terminus of **C**.

The magnitudes of the two components are known in Fig. 2-6*d*. A solution can be found by determining the intersection of arcs drawn from the origin and terminus of **C**, one of radius A and one of radius B. As the illustration shows, there are two solutions. Fortunately in a mechanism analysis it is usually clear which solution is wanted.

2-3. Motion

Definitions. *Motion* is a changing of position. *Displacement* is the change in position; it is, as Fig. 2-1 illustrates, a vector quantity. The time rate of change of position is *velocity*, similarly a vector quantity. When it is only the magnitude of a velocity that is of interest, it is referred to as *speed*. The time rate of change in velocity is *acceleration*. Unfortunately there is no generally accepted name for the magnitude of acceleration, though *magceleration* has been suggested. The time rate of change of acceleration is called *pulse or jerk*. No generally accepted name for the magnitude of pulse or jerk is available.

Types of motion. In kinematic analysis both the motions of particles and those of rigid bodies are of interest. The motion of a particle is necessarily linear, since it is not meaningful to speak of the rotation of a point. If a particle is moving on a straight line, it is said to have *rectilinear motion*; otherwise it has *curvilinear motion*.

When a rigid body moves in such a way that each particle has the same motion, the motion is called *translation*. If each particle is, in addition, moving in a straight line, it is *rectilinear translation*; otherwise it is *curvilinear translation*.

The type of motion most frequently of interest in mechanism study is plane motion. A body is said to have *plane motion* when its particles move in parallel or coincident planes. If a body has plane motion, it may be translation, rotation, or a combination of the two. In *plane rotation* the path of each particle is a circle whose center is on a fixed axis perpendicular to the plane of rotation.

The various types of plane motion are illustrated by the motions of the links of the double-crank and slider mechanism in Fig. 2-7. Links 2 and 4 have pure rotational motion while the motion of link 6 is one of pure translation. Link 3 also has a motion of translation. But while link 6 has rectilinear translation, the motion of link 3 is curvilinear translation. Link 5 has a plane motion which is a combination of rotation and translation.

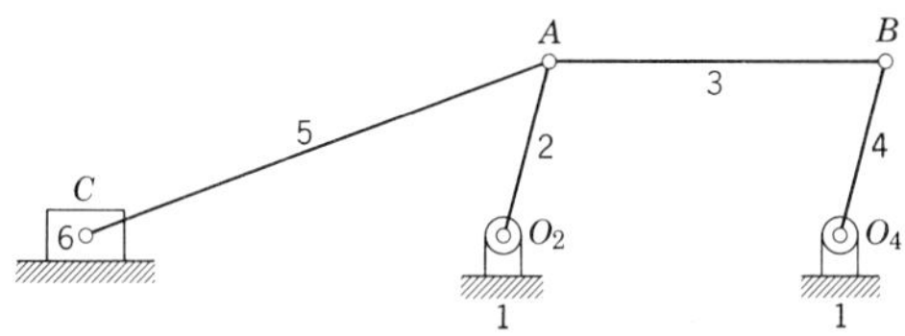

Fig. 2-7. Double crank and slider mechanism.

2-4. Rectilinear Motion

Motion formulas will be developed in this section for the motion of a particle along a straight line with uniform acceleration. In many practical problems a rigid body may be treated as a particle, if any rotation it may have is of no interest. For example, in many astronomical calculations the planets are regarded as particles with curvilinear motions.

In developing the equations for rectilinear motion, the vector notation will not be used, for keeping track of directions is exceptionally simple. A particle moving on a straight line can have but one of two directions. One of these directions can be regarded as positive and the other as negative.

Since acceleration is assumed constant, it can be found as the quotient of the change in velocity and the time interval in which the change took place. That is,

$$a = \frac{v - v_o}{t} \tag{2-1}$$

The displacement that occurs in any time interval is the product of the average velocity and the time interval.

$$s = \frac{v_o + v}{2} t \tag{2-2}$$

From Eqs. 2-1 and 2-2 two more equations can be derived which are useful for many problems. Although they add no new knowledge, they often simplify calculations.

Solving Eq. 2-1 for v,

$$v = v_o + at$$

Substituting for v in Eq. 2-2,

$$s = v_o t + \frac{at^2}{2} \tag{2-3}$$

Solving Eq. 2-1 for t,

$$t = \frac{v - v_o}{a}$$

Substituting for t in Eq. 2-2,

$$s = \frac{v^2 - v_o^2}{2a}$$

or

$$v^2 - v_o^2 = 2as \tag{2-4}$$

In Eqs. 2-1 to 2-4,

s = displacement, ft or in.
t = time interval, sec
v_o = initial velocity, ft/sec or in./sec
v = velocity at time t, ft/sec or in./sec
a = acceleration, ft/sec² or in./sec²

Example 2-1. A bolt was dropped from a blimp, 12,800 ft above the ground. The initial vertical velocity was zero. How long did it take the bolt to reach the ground?

Solution. For this problem, Eq. 2-3 will be useful. Since the initial velocity, v_o, is zero, Eq. 2-3 becomes

$$s = \frac{at^2}{2}$$

Solving for t,

$$t = \sqrt{\frac{2s}{a}}$$

For a falling body, the acceleration is that of gravity, 32.2 ft/sec². Substituting the data,

$$t = \sqrt{\frac{(2)(12{,}800)}{32.2}}$$

$$= 28.2 \text{ sec}$$

It will take the bolt 28.2 sec to reach the ground.

Example 2-2. A wagon starting from rest accelerated at a uniform rate to a maximum velocity, v, while traveling 18 ft. Then it decelerated at a uniform rate to rest while traveling another 12 ft. The total time was 16 sec. Find the acceleration rate, the maximum velocity reached, the deceleration rate, and the time taken for each part of the movement.

Solution

1. It must be recognized that there are two separate constant acceleration

intervals involved. Equations must be set up for each interval and solved simultaneously. (See Table 2-1.)

TABLE 2-1

Quantity	Interval 1	Interval 2
Initial velocity	0	v
Final velocity	v	0
Displacement	18 ft	12 ft
Acceleration	a_1	a_2
Time interval	t_1	t_2

2. Applying Eq. 2-2 to the first interval,

$$18 = \frac{0 + v}{2} t_1$$

and to the second interval,

$$12 = \frac{v + 0}{2} t_2$$

Solving each of these equations for v,

$$v = \frac{36}{t_1}$$

and

$$v = \frac{24}{t_2}$$

3. Therefore,

$$\frac{36}{t_1} = \frac{24}{t_2}$$

or

$$t_1 = 1.5t_2$$

4. But the sum of the two time intervals is given as 16 sec. That is,

$$t_1 + t_2 = 16$$

Solving these two equations for t_1 and t_2 simultaneously,

$$t_1 = 9.6 \text{ sec}$$

$$t_2 = 6.4 \text{ sec}$$

5. The value of v, the maximum velocity reached, can now be found.

$$v = \frac{36}{t_1}$$

$$= \frac{36}{9.6}$$

$$= 3.75 \text{ ft/sec}$$

6. The value of the acceleration for each interval can be found by using Eq. 2-1. For the first interval,

$$a_1 = \frac{3.75 - 0}{9.6}$$

$$= 0.391 \text{ ft/sec}^2$$

and for the second interval,

$$a_2 = \frac{0 - 3.75}{6.4}$$

$$= -0.586 \text{ ft/sec}^2$$

The negative sign for a_2 indicates a deceleration. During the second interval the wagon was slowing down.

2-5. Rotation

The angular displacement of a body can be measured in degrees, radians, or revolutions. A *radian* is the central angle subtended by a circular arc whose length is the same as the radius of the circle. Since the radius of a circle can be laid along its circumference 2π times, there are 2π radians in one revolution or 360°. Therefore,

$$\pi \text{ radians} = 180° \tag{2-5}$$

Angular velocities in this text usually will be specified either as radians per second or as revolutions per minute, though revolutions per second will also sometimes be used. To convert from revolutions per minute to radians per second, the following formula is used.

$$\omega = \frac{2\pi N}{60} \tag{2-6}$$

As in the previous article, vector notation will not be used in the equations for plane rotation. To indicate counterclockwise rotation, a positive sign will be used. This convention is adopted to obtain consistency with the conventions of trigonometry where positive angles are measured in a counterclockwise direction. A clockwise rotation will be considered negative.

If we regard angular acceleration as uniform, it is the quotient of the change in angular velocity and the time interval in which the change took place.

$$\alpha = \frac{\omega - \omega_o}{t} \tag{2-7}$$

The angular displacement is the product of the average angular velocity and the time interval.

$$\theta = \frac{\omega + \omega_o}{2} t \tag{2-8}$$

From these two basic equations, two more can be derived. Solving Eq. 2-7 for ω,

$$\omega = \omega_o + \alpha t$$

Substituting for ω in Eq. 2-8,

$$\theta = \omega_o t + \frac{\alpha t^2}{2} \tag{2-9}$$

Solving Eq. 2-7 for t,

$$t = \frac{\omega - \omega_o}{\alpha}$$

Substituting for t in Eq. 2-8,

$$\omega^2 - \omega_o^2 = 2\alpha\theta \tag{2-10}$$

The rotation of a body with constant angular velocity is a special case of constant angular acceleration. Substituting $\alpha = 0$ and $\omega = \omega_o$ in Eq. 2-9,

$$\theta = \omega t \tag{2-11}$$

In Eqs. 2-6 to 2-11,

θ = angular displacement, rad
t = time interval, sec
ω_o = initial angular velocity, rad/sec
ω = angular velocity at time t, rad/sec or average angular velocity, rad/sec
N = angular velocity, rpm
α = angular acceleration, rad/sec^2

Example 2-3. An automobile wheel initially turning at 400 rpm is accelerated in 10 sec to 865 rpm. Find the angular acceleration rate and the number of turns made by the wheel while it is being accelerated.

Solution. Applying Eq. 2-6,

$$\omega_o = \frac{2\pi(400)}{60}$$

$$= 42.0 \text{ rad/sec}$$

$$\omega = \frac{2\pi(865)}{60}$$

$$= 90.5 \text{ rad/sec}$$

Applying Eq. 2-7,

$$\alpha = \frac{90.5 - 42.0}{10}$$

$$= 4.85 \text{ rad/sec}^2$$

And, finally, applying Eq. 2-8,

$$\theta = \frac{90.5 + 42.0}{2} 10$$

$$= 662.5 \text{ rad or } 105.2 \text{ rev}$$

2-6. Motion of a Point on a Circular Path

Velocity. Figure 2-8 shows a body rotating about a point O. On the body is a point P at distance r from the center of rotation, O. The velocity

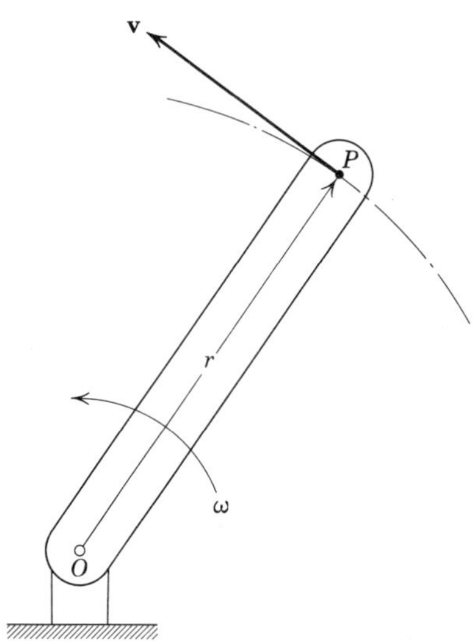

Fig. 2-8. Tangential velocity.

of P is tangential to the circular path of the point and thus perpendicular to the radius joining P to O. Since the velocity of P is tangential to the path of P, it is called the *tangential velocity*. Its sense is consistent with the sense of the angular velocity of the body carrying point P. Its magnitude is

$$v = r\omega \tag{2-12}$$

Equation 2-12 holds whether the angular velocity, ω, is constant or not. The equation gives the value of the tangential velocity for any instantaneous value of ω.

If the angular velocity is given in revolutions per minute, it is useful to have an equation for tangential velocity in terms of rpm. Substituting for ω from Eq. 2-6 in Eq. 2-12,

$$v = \frac{2\pi r N}{60} \tag{2-13}$$

While no formal proof of either Eq. 2-12 or 2-13 will be given here, the reader may examine Eq. 2-13 and convince himself in a rough, intuitive way of its validity. If a point is moving on a circle, it travels the circumference, $2\pi r$, each revolution. If it travels N revolutions per minute, then the distance covered is $2\pi rN$ per minute or $2\pi rN/60$ per second.
In Eqs. 2-12 and 2-13,

r = radius to point, ft or in.
ω = angular velocity, rad/sec
N = angular velocity, rpm
v = tangential velocity, ft/sec or in./sec

Acceleration. To find the acceleration of a point P moving on a circular path (Fig. 2-9), the tangential and normal components are found separately and their resultant determined.

If a point is moving along a circular path, its velocity is constantly changing direction, even when the angular velocity, and hence the magnitude of the tangential velocity, is constant. That component of P's acceleration which expresses the change in the direction of the tangential velocity is the *normal acceleration*. Its sense is always toward the center of rotation and its magnitude is

$$a_n = r\omega^2 \tag{2-14}$$

Solving Eq. 2-12 for ω,

$$\omega = \frac{v}{r}$$

and substituting in Eq. 2-14,

$$a_n = \frac{v^2}{r} \tag{2-14a}$$

Equation 2-14 can also be written as

$$a_n = (r\omega)\omega$$

Substituting for $r\omega$ from Eq. 2-12,

$$a_n = v\omega \tag{2-14b}$$

Three equations are available for the calculation of the normal acceleration. The choice between them is a matter of convenience.

If the angular velocity is not constant, that is, if there is an angular acceleration, the magnitude of the tangential velocity will also be changing. That acceleration component which expresses the change in magnitude is

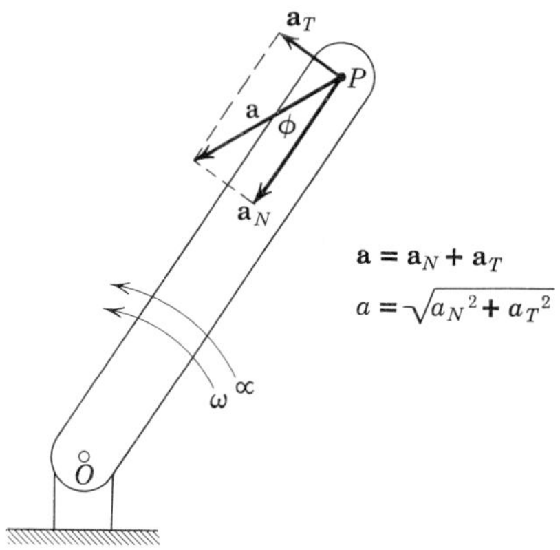

Fig. 2-9. Acceleration of a point on a circular path.

the *tangential acceleration.* It is always tangent to the circular path, and thus perpendicular to the radius, and its sense is consistent with that of the angular acceleration. The magnitude of the tangential acceleration is

$$a_t = r\alpha \tag{2-15}$$

Once the normal and tangential components of the acceleration of point P have been found, the resultant acceleration is given by the vector equation:

$$\mathbf{a} = \mathbf{a}_n + \mathbf{a}_t \tag{2-16}$$

Since these components are mutually perpendicular, the magnitude of the resultant can be found by an application of the Pythagorean Theorem.

$$a = \sqrt{a_n^2 + a_t^2} \tag{2-17}$$

The position of the total acceleration vector is described by angle ϕ (Fig. 2-9), the angle between the acceleration and the radius to the point. The angle ϕ is

$$\phi = \text{arc tan}\,\frac{a_t}{a_n} \tag{2-18}$$

In Eqs. 2-12 to 2-18,

α = angular acceleration, rad/sec^2
a_n = magnitude of normal acceleration, ft/sec^2 or in./sec^2
a_t = magnitude of tangential acceleration, ft/sec^2 or in./sec^2
a = magnitude of resultant acceleration, ft/sec^2 or in./sec^2
ϕ = angle between resultant acceleration and radius, degrees

Example 2-4. A small precision gear in a control mechanism is accelerated from rest to 3000 rpm in 3 sec. What is the total acceleration of a point 2 in. from the center of the gear at the end of the first half second?

Solution

1. Applying Eq. 2-7 to the 3-sec interval,

$$\begin{aligned}\alpha &= \frac{\omega - \omega_o}{t} \\ &= \frac{3000(2\pi/60) - 0}{3} \\ &= 104.7 \text{ rad/sec}^2\end{aligned}$$

2. Applying Eq. 2-7 again, but this time to the 0.5-sec interval to find the instantaneous angular velocity at $t = 0.5$ sec,

$$\begin{aligned}\omega &= \omega_o + \alpha t \\ &= 0 + (104.7)(0.5) \\ &= 52.35 \text{ rad/sec}\end{aligned}$$

3. The magnitudes of the normal and tangential components can now be found at $t = 0.5$ sec. Using Eq. 2-14,

$$\begin{aligned}a_n &= r\omega^2 \\ &= (2/12)(52.35)^2 \\ &= 456 \text{ ft/sec}^2\end{aligned}$$

and using Eq. 2-15,

$$\begin{aligned}a_t &= r\omega \\ &= (2/12)(104.7) \\ &= 17.45 \text{ ft/sec}^2\end{aligned}$$

4. Now that the tangential and normal components have been found, the resultant acceleration can be determined. Employing Eq. 2-17,

$$\begin{aligned}a &= \sqrt{a_n^2 + a_t^2} \\ &= \sqrt{(456)^2 + (17.45)^2} \\ &= 456.3 \text{ ft/sec}^2\end{aligned}$$

5. And, finally, using Eq. 2-18,

$$\phi = \text{arc tan}\,\frac{a_t}{a_n}$$

$$= \text{arc tan}\,\frac{17.45}{456}$$

$$\phi = 2^\circ\,12'$$

2-7. Rolling Wheel

Figure 2-10 shows a wheel of diameter D rolling on a flat surface with an angular speed of N. The magnitude of the velocity of the wheel's center, or axle speed, is given by the following equation:

$$V = \frac{\pi DN}{60} \tag{2-19}$$

where V = axle speed, ft/sec or in./sec
D = wheel diameter, ft or in.
N = wheel speed, rpm

Equation 2-19 can be rewritten as

$$V = R\omega \tag{2-20}$$

where R = wheel radius, ft or in.
ω = wheel speed, rad/sec

This last equation has the same form and units as Eq. 2-12 although its significance is different.

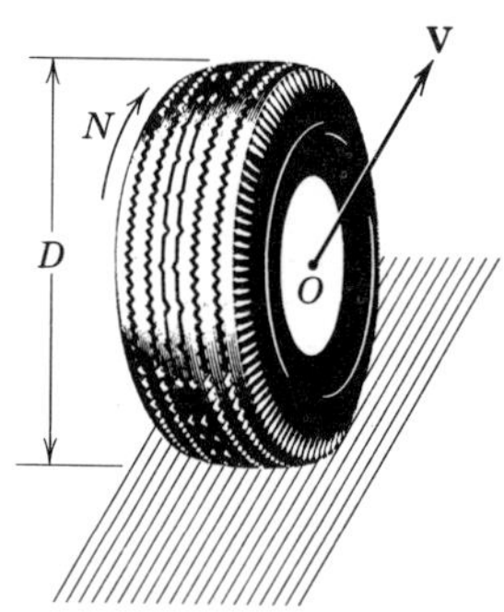

Fig. 2-10. Rolling wheel.

Example 2-5. What is the angular speed of an automobile wheel of 30 in. diameter if the vehicle is traveling 50 mph?

Solution

1. Solving Eq. 2-19 for N,

$$N = \frac{60V}{\pi D}$$

2. For convenience the car speed will first be converted from miles per hour to feet per second. Since 60 mph and 88 ft/sec are equivalent,

$$\frac{V}{88} = \frac{50}{60}$$

$$V = 73.4 \text{ ft/sec}$$

3. Solving for N,

$$N = \frac{(60)(73.4)}{(\pi)(30/12)}$$

$$= 562 \text{ rpm}$$

2-8. Graphical Scales

The use of graphical techniques to solve motion problems requires the representation of the magnitudes of vector quantities by line segments. Graphical scales are also essential for the representation of mechanisms which are either too large or too small to be represented conveniently at full size.

Definition. In this text scales will be specified as ratios. A *scale* is defined as

$$\text{scale} = \frac{\text{quantity represented}}{\text{drawing distance}} \tag{2-21}$$

For example, suppose that on an architect's drawing a building dimension which is actually 22 ft is represented by a distance on the drawing which measures 1.1 in. Then the distance scale used in preparing the drawing must have been, using Eq. 2-21,

$$k_s = \frac{22 \text{ ft}}{1.1 \text{ in.}}$$

$$= 20 \text{ ft/in.}$$

That is, each inch on the drawing represents 20 ft.

Scale layout. In kinematic analysis it will sometimes happen that the graph representing the variation in some motion quantity such as velocity will be constructed by a short-cut graphical technique, rather than by point-by-point plotting from a table of calculated data. In such cases, to make the graph readable after it is constructed, it is necessary to lay out scale divisions on the axes to a scale dictated by the construction technique. The scale marks must be laid out in such a way that points which fall between scale marks can easily be interpolated by eye. The scale divisions should represent quantities such as 1, 2, 5, 10, rather than 3, 7. An example will illustrate the problem and the method of dealing with it.

Example 2-6. The y-axis of a graph is to represent velocities to the scale 37.8 ft/sec/in. Velocities in the range 0 to 150 ft/sec must be represented. Decide upon appropriate scale divisions and determine the drawing distance between scale marks.

Solution

1. The scale division to be used is a matter of decision. However, it should be one such that interpolation by eye is possible. Scale divisions representing 10 ft/sec will be used.

2. The remaining problem is to calculate the drawing distance between scale marks. From Eq. 2-21,

$$\text{drawing distance} = \frac{\text{quantity represented}}{\text{scale}}$$

$$= \frac{10 \text{ ft/sec}}{37.8 \text{ ft/ sec/in.}}$$

$$= 0.265 \text{ in.}$$

The scale marks should be placed 0.265 in. apart. In practice, if each scale mark is measured off from the previous one, cumulative error is likely to be great. Therefore each scale mark should be laid out from the origin. For example, the distance from the origin to the scale mark representing 110 ft/sec will be

$$\text{drawing distance} = \frac{110 \text{ ft/sec}}{37.8 \text{ ft/sec/in.}}$$

$$= 2.92 \text{ in.}$$

This technique involves more work, but it avoids cumulative error.

2-9. Relative Motion

One of the most powerful and versatile methods of motion analysis is the relative motion approach. Only the fundamentals of this method will be studied here; it will be developed in more detail in Chapters 3 and 4, where it will be applied to the motion analyses of various mechanisms.

Relative displacement. The displacements of two points, A and B, relative to a third point O are shown in Fig. 2-11. The displacement of point A is $\mathbf{s}_A$, and that of B is $\mathbf{s}_B$. The displacement of A relative to B, $\mathbf{s}_{A/B}$ has also been drawn in as a vector from B and toward A. When

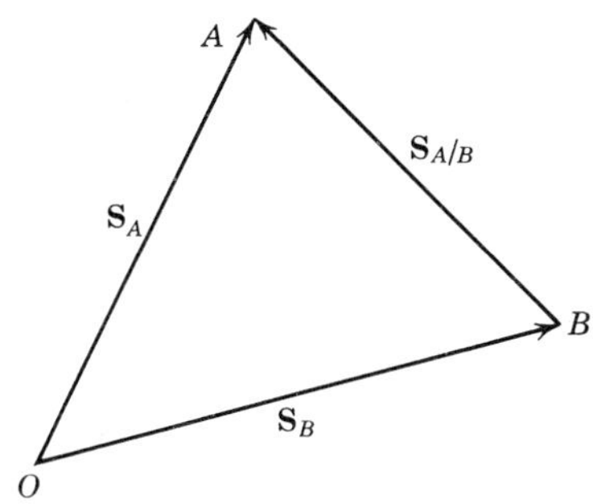

$\mathbf{s}_{A/B} = \mathbf{s}_A - \mathbf{s}_B$
Fig. 2-11. Relative displacement.

recalling the techniques of vector subtraction and inspecting the figure, it can be seen that

$$\mathbf{s}_{A/B} = \mathbf{s}_A - \mathbf{s}_B \tag{2-22}$$

This equation expresses the general rule:

> The displacement of one point relative to a second is equal to the displacement of the first minus the displacement of the second, both relative to some third point.

Equation 2-22 can be rewritten as

$$\mathbf{s}_A = \mathbf{s}_B + \mathbf{s}_{A/B} \tag{2-23}$$

This equation expresses the rule:

> The displacement of a point is equal to the displacement of a second point plus the displacement of the first relative to the second.

Example 2-7. A weather balloon is 13 miles N 50° E of Logan International Airport. A plane is 20 miles N 19° W of the airport. What is the displacement of the balloon relative to the plane?

Solution. The solution is in Fig. 2-12. The airport is A, the balloon B, and the plane P. Applying Eq. 2-22,

$$\mathbf{s}_{B/P} = \mathbf{s}_B - \mathbf{s}_P$$

The displacements $\mathbf{s}_B$ and $\mathbf{s}_P$ are each relative to the airport and are given. They have been laid out in Fig. 2-12 with their origins coincident. Their vector difference, $\mathbf{s}_{B/P}$, is the vector from the terminus of $\mathbf{s}_P$ to that of $\mathbf{s}_B$. The result is

$$\mathbf{s}_{B/P} = 19.6 \text{ miles S } 57.5° \text{ E}$$

Relative velocity. The analysis for relative velocity follows a similar pattern. For convenience the velocity of a point on a link relative to the

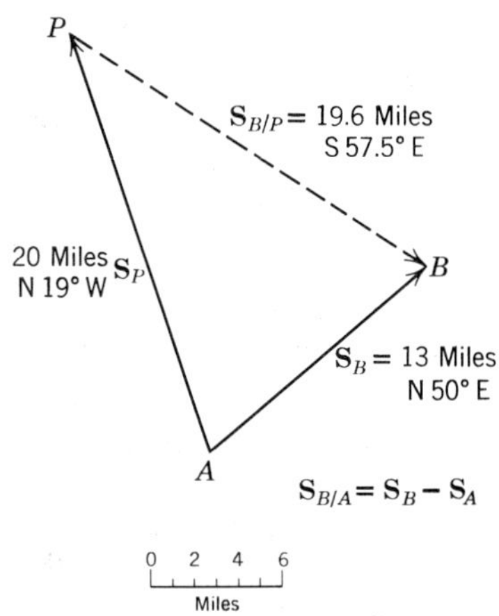

Fig. 2-12. Solution of Example 2-7.

frame of the mechanism will be called its *absolute velocity*. Its velocity relative to another moving point will be a *relative velocity*.

The relative velocity of two points in a mechanism is given by the vector equation,

$$\mathbf{v}_{A/B} = \mathbf{v}_A - \mathbf{v}_B \tag{2-24}$$

which expresses the rule:

> For any two moving points, *A* and *B*, the velocity of *A* relative to *B* is the absolute velocity of *A* minus that of *B*.

Equation 2-24 can be rewritten as

$$\mathbf{v}_A = \mathbf{v}_B + \mathbf{v}_{A/B} \tag{2-25}$$

which expresses the rule:

> The velocity of one point is the sum of the velocity of any second point and the velocity of the first relative to the second.

Example 2-8. Figure 2-13*a* shows a belt and pulley drive. Pulley 2 is 18 in. in diameter and is rotating at 400 rpm clockwise. It is belted to pulley 3, which has a diameter of 9 in. Determine the velocity of point *A* on pulley 2 relative to point *B* on pulley 3.

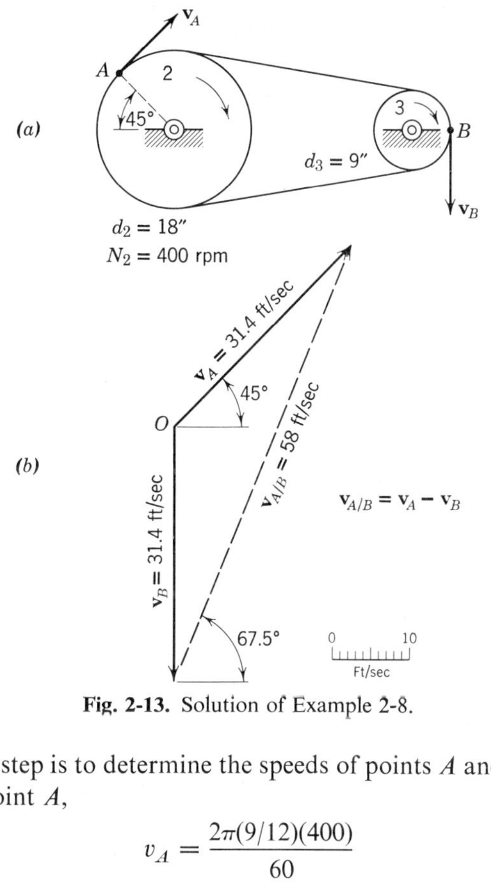

Fig. 2-13. Solution of Example 2-8.

Solution

1. The first step is to determine the speeds of points *A* and *B*. Applying Eq. 2-13 to point *A*,

$$v_A = \frac{2\pi(9/12)(400)}{60}$$

$$= 31.4 \text{ ft/sec}$$

If we assume no slip between the belt and the pulleys, the speed of *B* must be the same as that of *A*.

2. The directions of the velocities of *A* and *B* are determined by inspection and have been indicated in Fig. 2-13*a*.

3. In Fig. 2-13*b* velocities $\mathbf{v}_A$ and $\mathbf{v}_B$ have been drawn to scale and with a common origin. The relative velocity, $\mathbf{v}_{A/B}$, is the vector from the tip of $\mathbf{v}_B$ to that of $\mathbf{v}_A$. The result is

$$\mathbf{v}_{A/B} = 58.0 \text{ ft/sec at } 67.5°$$

Relative velocity of two points on a rigid link. A special case of relative velocity of great importance is that of the relative velocity of two points on the same rigid body.

A body is said to be *rigid* if the distance between any two points on the body remains constant. In Fig. 2-14 a rigid body m undergoing plane motion is shown. At an instant the velocities of two points, A and B, on

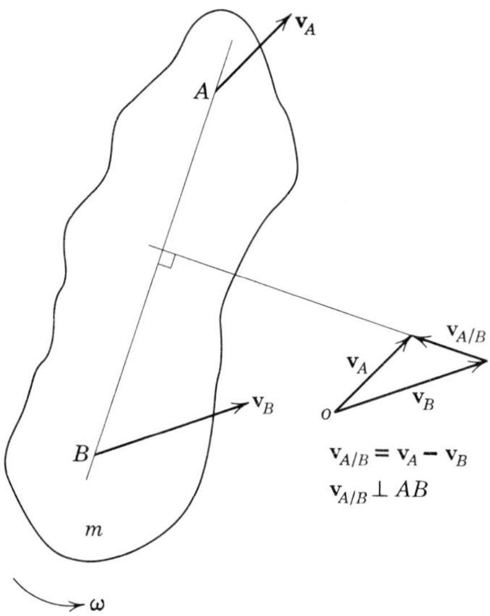

Fig. 2-14. Relative velocity of two points on a rigid body.

the body are shown. Since distance AB is constant, the relative velocity of A to B must have no component along the line joining A and B. If it did, this would imply that A and B are either approaching each other or receding from each other. Either implication is contrary to the assumption that the body is rigid. Therefore the following rule is justified:

> The relative velocity of any two points on the same rigid body is perpendicular to the line joining the two points.

When we speak of the velocity of one point on a rigid body relative to another, say A relative to B, the movement of B *relative* to A is necessarily on a circular arc of radius AB and with center at A. The relative velocity of A to B is the velocity A would have relative to ground if B were fixed and the body were rotating about B with its angular velocity ω. Its magnitude is, by Eq. 2-12,

$$v_{A/B} = AB \cdot \omega \tag{2-26}$$

This can be rewritten as

$$\omega = \frac{v_{A/B}}{AB} \tag{2-27}$$

which justifies the rule:

> The angular speed of a rigid body having plane motion can be found by dividing the magnitude of the relative velocity of any two points on the body by the distance between them.

This rule will be important in determining the angular velocities of links in a mechanism once the velocities of key points in the mechanism have been found.

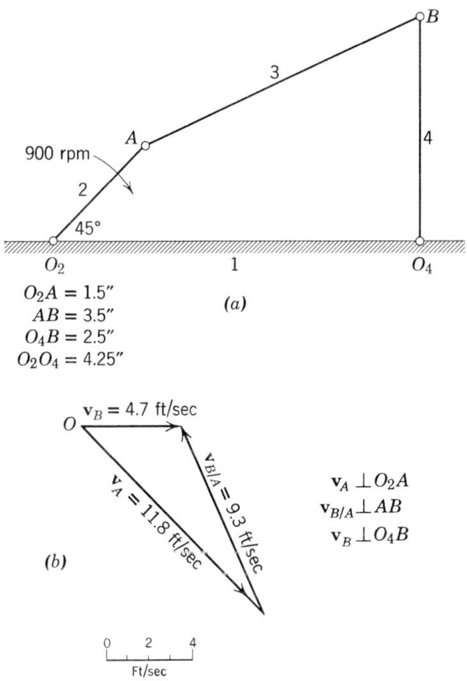

Fig. 2-15. Solution of Example 2-9.

Example 2-9. The crank of a four-bar linkage, illustrated in Fig. 2-15*a* is rotating clockwise at 900 rpm. The dimensions of the links and the angular position of the crank are shown on the drawing. Determine the velocity of point *B* and the angular velocity of link 3.

Solution

1. The first step is to calculate the speed of point *A*, using Eq. 2-13.

$$v_A = \frac{2\pi(1.5/12)(900)}{60}$$

$$= 11.8 \text{ ft/sec}$$

2. Next, the velocity of point B can be found, using Eq. 2-25:

$$\mathbf{v}_B = \mathbf{v}_A + \mathbf{v}_{B/A}$$

In Fig. 2-15b $\mathbf{v}_A$ has been laid out to scale, perpendicular to link 2, and with a sense consistent with that of the angular motion of link 2. If we use the triangle method of vector addition, the vector $\mathbf{v}_{B/A}$ must be placed with its origin at the terminus of $\mathbf{v}_A$. Though its magnitude is unknown, its direction is: it is perpendicular to the line joining points A and B. It is drawn of indefinite length from the terminus of $\mathbf{v}_A$. The magnitude of $\mathbf{v}_B$ also is unknown, but not its direction. It must be perpendicular to link 4 since B is on link 4, which rotates about a fixed pivot. To close the vector triangle, $\mathbf{v}_B$ must pass through the origin of $\mathbf{v}_A$. Thus a line of indefinite length is drawn in the direction of $\mathbf{v}_B$ and through the origin of $\mathbf{v}_A$. The intersection of the two lines just drawn completes the vector triangle.

3. To find the angular speed of link 3, Eq. 2-27 can be applied.

$$\begin{aligned}\omega_3 &= \frac{V_{B/A}}{BA} \\ &= \frac{9.3 \text{ ft/sec}}{3.5/12 \text{ ft}} \\ &= 31.9 \text{ rad/sec}\end{aligned}$$

Since the velocity of B relative to A is up and to the left, the motion of link 3 must be counterclockwise.

2-10. Simple Harmonic Motion

Definitions. When motion repeats itself at regular intervals it is said to be *periodic*. An especially simple kind of periodic motion is that called *simple harmonic motion*. This is the motion of a point moving on a straight line in such a way that its acceleration is always proportional to its displacement from the starting point and always toward the starting point.

Perhaps an easier way to visualize this motion is to refer to Fig. 2-16. Here a line, OP, is rotating about point O at a constant angular velocity. Point P' is the projection of P on the horizontal diameter DD'. As P moves in its circular path, its projection P' moves along the diameter DD'. Point P' has simple harmonic motion. Line OP will be called the *generating arm* and point P the *generating point*.

A mechanism that produces simple harmonic motion is the Scotch yoke, shown in Fig. 2-17. Crank OP corresponds to the generating arm and the yoke is the projection point, P'. As the arm turns at a uniform rate, the yoke reciprocates with simple harmonic motion.

If a weight is hung on an ordinary helical tension spring, given a small displacement, and released, the weight will vibrate up and down with simple harmonic motion. The motion of a pendulum with a fairly long cord is approximately simple harmonic.

In Fig. 2-16 the position of the generating arm is defined by the angle θ. The angular position of the arm is called its *phase*. The *amplitude* of the motion is the maximum distance point P' travels from origin O. It is equal to the radius of the generating arm, r.

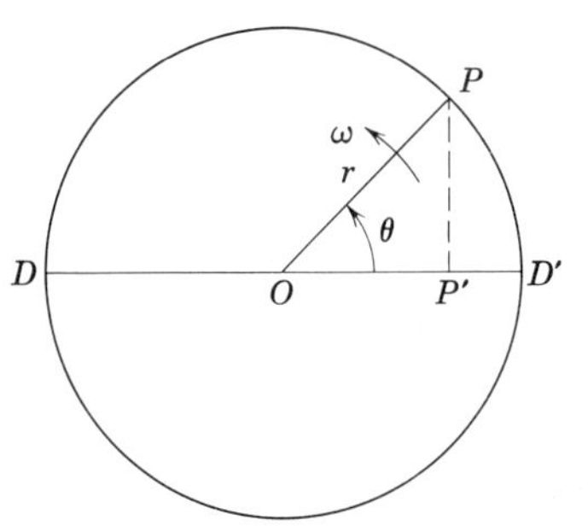

Fig. 2-16. Simple harmonic motion.

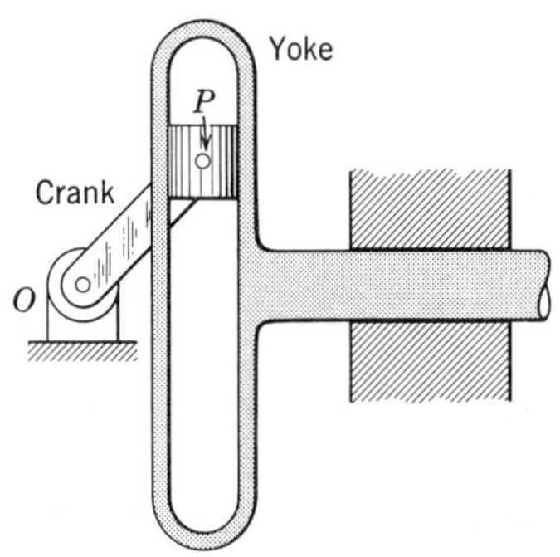

Fig. 2-17. Scotch yoke mechanism.

The time required for one complete motion cycle is called the *period*. In simple harmonic motion this is the time required for the generating arm to complete one revolution. Since the generating arm moves with constant angular velocity, Eq. 2-11 applies. Solving this equation for t,

$$t = \frac{\theta}{\omega}$$

The angular displacement of the generating arm corresponding to the period, T, is a full revolution or 2π radians. Hence,

$$T = \frac{2\pi}{\omega} \tag{2-28}$$

If the generating arm velocity is given in revolutions per minute, the above equation can be rewritten by substituting from Eq. 2-6 for ω. The result is

$$T = \frac{60}{N} \tag{2-29}$$

In Eqs. 2-28 and 2-29,

T = period, sec
ω = angular speed of generating arm, rad/sec
N = angular speed of generating arm, rpm

Displacement. The displacement of a point having simple harmonic motion is measured from the center of its path, O, Fig. 2-18*a*, which is also the center of rotation of the generating arm. When to the right of O, the

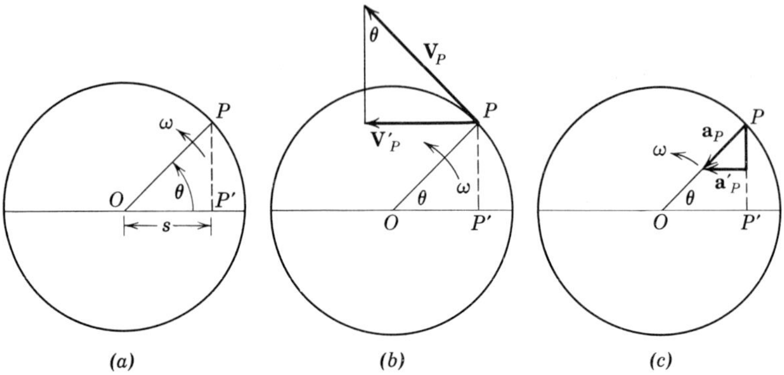

Fig. 2-18. Displacement, velocity, and acceleration in simple harmonic motion.

displacement, s, will be regarded as positive; when to the left, negative. The displacement is given by the equation

$$s = r \cos \theta \tag{2-30}$$

This can be rewritten by substituting for θ from Eq. 2-11,

$$s = r \cos \omega t \tag{2-30a}$$

where s = displacement, ft or in.
r = length of generating arm, ft or in.
θ = phase, degrees or radians
ω = angular speed of generating arm, rad/sec
t = time, sec

Velocity. When P' is moving toward the right, in the direction of increasing displacement, its velocity is positive; when moving to the left, its velocity is negative.

Since P' is the projection of P on the horizontal diameter of the generating circle, its velocity must be the same as the horizontal component of the velocity of the generating point P, otherwise P' would not remain directly under P. In Fig. 2-18*b* the velocity of the generating point, v_P, is shown; and it has been resolved into horizontal and vertical components. When we apply Eq. 2-12, the velocity of P is

$$v_P = r\omega$$

Its horizontal component, which is the velocity of P', is

$$v = -r\omega \sin \theta \qquad (2\text{-}31)$$

which can be rewritten, using Eq. 2-11, as

$$v = -r\omega \sin \omega t \qquad (2\text{-}31a)$$

where v = velocity of point having simple harmonic motion, ft/sec or in./sec
r = length of generating arm, ft or in.
ω = angular velocity of generating arm, rad/sec, positive if ccw
θ = phase, radians or degrees
t = time, sec

The negative signs have been inserted in Eqs. 2-31 and 2-31*a* to make them consistent with the sign conventions which have been adopted. For example, if the generating is moving counterclockwise but is in the first quadrant, the velocity should be negative, for the point P' will be moving to the left. Since both ω and sin θ will be positive under these conditions, and since r is always positive, the equation will result in a negative value for v.

Acceleration. Acceleration, like displacement and velocity, will be regarded as positive if toward the right and negative if toward the left.

The acceleration of P' must be the same as the horizontal component of P, if P' is to remain the projection of P. The acceleration of P will have only a normal component, since the angular velocity of the generating arm is constant. By using Eq. 2-14,

$$a_P = r\omega^2$$

Therefore the acceleration of P' is (Fig. 2-18*c*)

$$a = -r\omega^2 \cos \theta \qquad (2\text{-}32)$$

or, using Eq. 2-11,

$$a = -r\omega^2 \cos \omega t \qquad (2\text{-}32a)$$

Again negative signs have been inserted to make the equations consistent with the sign conventions.

Variation in displacement, velocity, and acceleration. The graphs in Fig. 2-19 illustrate the variation in these quantities for simple harmonic motion. Each of these quantities is either a function of the sine or of the cosine of the phase angle; hence each graph is sinusoidal in shape. The acceleration is numerically the greatest at the extreme ends of the path of the point, while the velocity is the greatest when the point is at the center of its path.

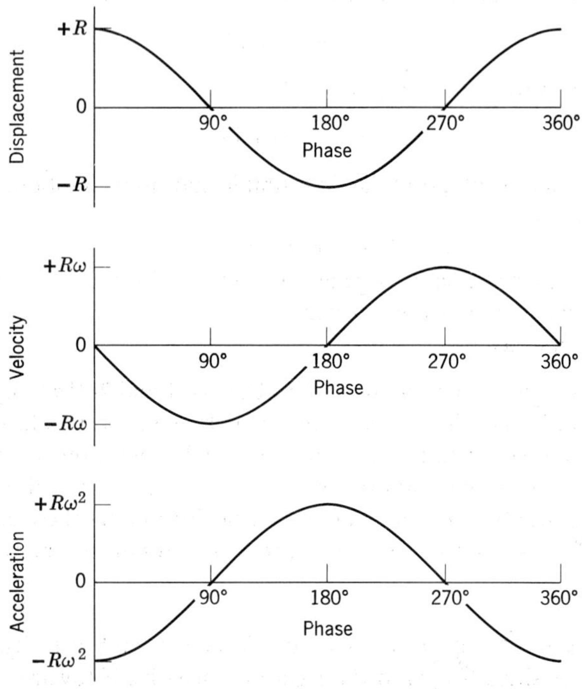

Fig. 2-19. Variation of displacement, velocity, acceleration in simple harmonic motion.

Example 2-10. A Scotch yoke mechanism has a 9.6 in. crank which turns at 600 rpm ccw. Determine: (1) the period; (2) the amplitude of the yoke's motion; (3) the phase at which acceleration is a maximum; and (4) the yoke displacement, velocity, and acceleration at $\theta = 30°, 115°, 240°$, and $315°$.

Solution

1. To find the period, Eq. 2-29 can be used.

$$T = \frac{60}{600}$$
$$= 0.1 \text{ sec}$$

2. The amplitude in simple harmonic motion is necessarily the radius of the crank. Therefore, in this instance it is 9.6 in. or 0.8 ft.

3. Inspection of Eq. 2-32 reveals the crank positions at which the yoke acceleration will be a maximum. Since both r and ω are constant, the acceleration will be numerically a maximum at those phases which make

$\cos\theta$ numerically a maximum. This will be at 0° or 180°. Inspection of the acceleration graph in Fig. 2-19 confirms this conclusion.

4. The displacement, acceleration, and velocity of the yoke can be found at the given phases by using Eqs. 2-30, 2-31, and 2-32. But first the constant quantities r, $r\omega$, and $r\omega^2$ must be evaluated:

$$\begin{aligned} r &= 9.6 \text{ in.} = 0.8 \text{ ft (given)} \\ r\omega &= 0.8(2\pi)(600)/60 \\ &= 49.86 \text{ ft/sec} \\ r\omega^2 &= (0.8)[(2\pi)(600)/60]^2 \\ &= 3158 \text{ ft/sec}^2 \end{aligned}$$

The calculations can now be made. They have been done in Table 2-2.

TABLE 2-2

θ deg.	$\sin\theta$	$\cos\theta$	$s = r\omega\cos\theta$ ft	$v = -r\omega\sin\theta$ ft/sec	$a = -r\omega^2\cos\theta$ ft/sec²
30	0.500	0.866	0.694	−24.9	−2735
115	0.966	−0.259	−0.207	−49.7	819
240	−0.866	−0.500	−0.400	43.2	1579
315	−0.707	0.707	0.565	35.2	−2263

2-11. Graphical Differentiation

Theory. Sometimes displacement-time data are available for a moving point in a mechanism, but data regarding the point's velocities and accelerations at various instants are lacking. For example, a motion picture might have been made, and from it displacement-time data obtained. To obtain the velocities and accelerations, the method of graphical differentiation can be used.

Graphical differentiation of motion curves for mechanisms is based on the following rules, which can be justified by the differential calculus:

1. The velocity of a point at any instant is proportional to the slope of the tangent to the displacement-time curve for that instant.
2. The acceleration of a point at any instant is proportional to the slope of the tangent to the velocity-time curve for that instant.
3. The pulse (or jerk) of a point at any instant is proportional to the slope of the tangent to the acceleration-time curve for that instant.

Velocity is the rate of change of displacement with respect to time and is called the *derivative* of displacement with respect to time. Similarly,

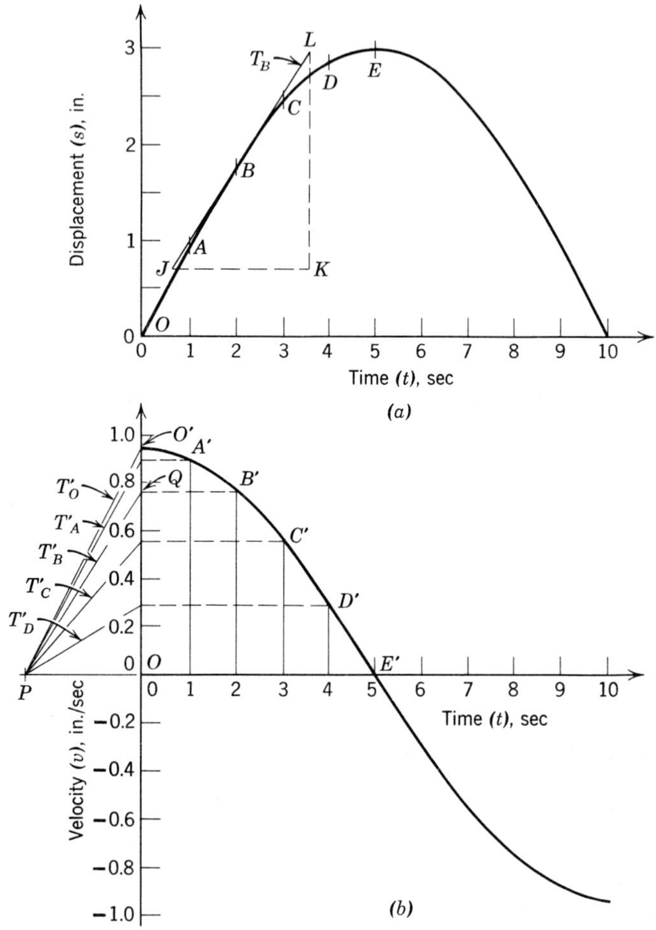

Fig. 2-20. Graphical differentiation.

acceleration is the derivative of velocity with respect to time, and pulse is the derivative of acceleration with respect to time. In general, for an equation

$$y = f(x)$$

the rate of change of y with respect to x is called the *derivative* of y with respect to x, written dy/dx, and at any point is proportional to the slope of the tangent to the graph of the equation. If y and x are plotted to the same scale, the derivative is equal to the slope.

In Fig. 2-20*a* line T_B has been drawn tangent to the displacement-time curve at point B ($t = 2$ sec). The velocity at B is proportional to the slope of this tangent line. Points J and L have been chosen on it arbitrarily, and from them lines drawn parallel to the time and displacement axes to form a right triangle. The velocity at B is equal to the quotient of the displacement *represented* by side LK divided by the time interval *represented* by side JK. That is,

$$V_B = \frac{\Delta s}{\Delta t} = \frac{LK \cdot k_s}{JK \cdot k_t} = \frac{k_s}{k_t}\left[\frac{LK}{JK}\right]$$

where V_B = the velocity at B
k_s = the displacement scale
k_t = the time scale
LK and JK = the actual drawing distances

The scales are constants and LK/JK is the slope of the tangent line. This illustrates the rule that the velocity at any instant is proportional to the slope of the line tangent to the displacement-time curve for that instant.

Technique. The technique is illustrated in Fig. 2-20.

1. Select convenient displacement and time scales, and plot the displacement-time curve. This has been done in Fig. 2-20*a*.
2. Select a convenient velocity scale and calculate the pole distance, PO, using Eq. 2-33. Locate the pole. For the example, this has been done in Fig. 2-20*b*.
3. Select points on the curve and at them carefully draw lines tangent to the curve. In the example, Fig. 2-20*a*, points O, A, B, C, D, etc. were selected, though the tangent T_B at B only has been shown.
4. From the pole, P, draw lines to the velocity axis parallel to these tangents. For example, T'_B (PQ) is parallel to T_B.
5. Project the points on the velocity axis thus found to the corresponding ordinates. For example, since T_B is the tangent at $t = 2$ sec, T'_B also applies at $t = 2$ sec. Therefore Q is projected to the ordinate for $t = 2$, locating point B'.
6. Using a french curve, connect points O', A', B', C', D', etc.

It remains to develop an equation for the pole distance, PO. Since

$$V_B = \frac{k_s}{k_t}\left(\frac{LK}{JK}\right)$$

and

$$V_B = OQ \cdot k_v$$

then

$$OQ \cdot k_v = \frac{k_s}{k_t}\left(\frac{LK}{JK}\right)$$

But since T_B and T'_B are drawn to the same slope,

$$\frac{LK}{JK} = \frac{OQ}{PO}$$

Hence

$$OQ \cdot k_v = \frac{k_s}{k_t}\left(\frac{OQ}{PO}\right)$$

or

$$PO = \frac{k_s}{k_v \cdot k_t} \tag{2-33}$$

Any three of the four quantities in Eq. 2-33 can be selected arbitrarily and the fourth calculated. For example, in the original drawing for Fig. 2-20 the scales were selected as $k_s = 1$ in./in., $k_t = 2$ sec/in. and $k_v = 0.4$ in./sec/in. Then the pole distance, PO, was found, using Eq. 2-33,

$$PO = \frac{1}{(2)(0.4)}$$
$$= 1.25 \text{ in.}$$

Distance PO could have been selected arbitrarily instead, and the velocity scale calculated.

The procedure for differentiating a velocity-time curve to determine acceleration is similar. Instead of using Eq. 2-33 for calculating the pole distance, PO, the following equation must be used:

$$PO = \frac{k_v}{k_a \cdot k_t} \tag{2-34}$$

And for the differentiation of an acceleration-time curve to obtain pulse,

$$PO = \frac{k_a}{k_p \cdot k_t} \tag{2-35}$$

In Eqs. 2-33 to 2-35,

PO = pole distance, in.
k_s = displacement scale, in./in. or ft/in.
k_t = time scale, sec/in.
k_v = velocity scale, in./sec/in. or ft/sec/in.
k_a = acceleration scale, in./sec^2/in. or ft/sec^2/in.
k_p = pulse scale, in./sec^3/in. or ft/sec^3/in.

Prediction of shape of derived curve. When differentiating a curve, we can predict the general shape of the derived curve from an inspection of the given curve, using the following rules:

1. At a maximum or minimum point on a curve the derivative is zero.
2. At an inflection point on the given curve the derivative will be numerically a maximum.
3. If the curve is rising, the derivative is positive; if falling, it is negative.

These rules can be illustrated by reference to Figs. 2-20 and 2-21. At point E in Fig. 2-20*a* the curve has reached a maximum point; a line drawn tangent to the curve at either a maximum or a minimum point will have zero slope. Hence at $t = 5$ the velocity is zero, as Fig. 2-20*b* shows.

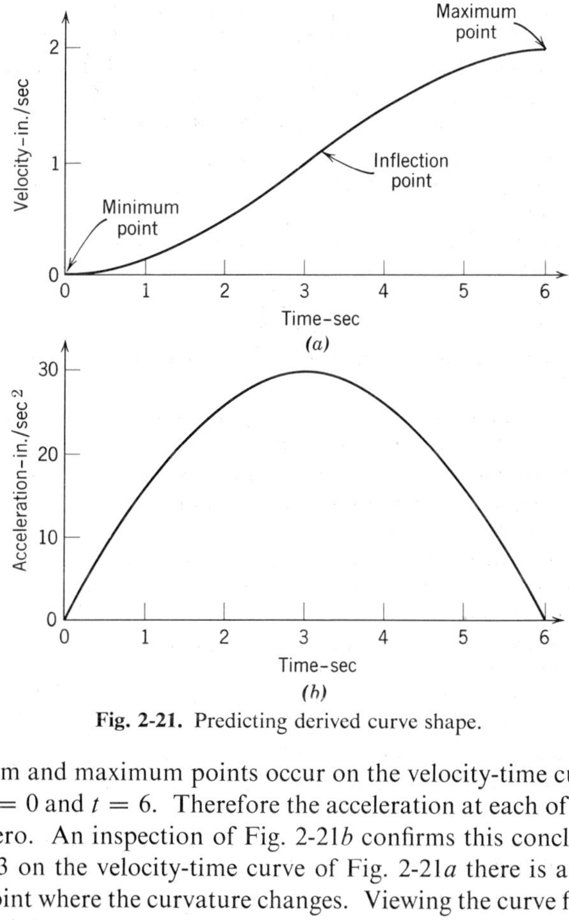

Fig. 2-21. Predicting derived curve shape.

Minimum and maximum points occur on the velocity-time curve of Fig. 2-21*a* at $t = 0$ and $t = 6$. Therefore the acceleration at each of these times must be zero. An inspection of Fig. 2-21*b* confirms this conclusion.

At $t = 3$ on the velocity-time curve of Fig. 2-21*a* there is an inflection point, a point where the curvature changes. Viewing the curve from above, we see that it is concave to the left of $t = 3$ and convex to the right. As the second of the above three rules specifies, the acceleration will reach a maximum value, numerically, at this point. The acceleration-time curve in Fig. 2-21*b* shows a crest at $t = 3$.

In Fig. 2-20*a* the displacement increases up to $t = 5$ and then decreases. To the left of this point a line tangent to the curve will fall in the first (or third) quadrant, thus having a positive slope. To the right of this point, a tangent will fall in the second (or fourth) quadrant and thus have a negative slope. Therefore, to the left of $t = 5$, the velocity will be positive and to the right negative. An inspection of the velocity-time curve in Fig. 2-20*b* confirms this reasoning.

PROBLEMS

2-1. A point is given a displacement of 2.7 in. at 19° followed by a second displacement of 1.9 in. at −22°. What is the displacement of the point relative to its starting point after these displacements? Solve graphically.

2-2. Force **A** is 42 lb at 21° and **B** is 61 lb at −65°. Find **A** + **B** graphically.

2-3. Find graphically the resultant of the following forces: **A** = 25 lb at 0°; **B** = 20 lb at 30°; **C** = 60 lb at 90°; and **D** = 50 lb at 225°.

2-4. An airship is traveling in a straight line at a velocity of 125 mph at N 53° W. Find the components of its velocity in the directions N 42° E and S 70° W. Solve graphically.

2-5. Find the North and East components of a velocity of 250 mph N 37° E. Solve both graphically and analytically.

2-6. Find the horizontal and vertical components of a force of 2000 lb at 23°. Use both graphical and analytical methods.

2-7. The velocity $\mathbf{V}_A$ is 75 mph N 16° W. The velocity $\mathbf{V}_B$ is 110 mph S 28° W. Find $\mathbf{V}_A - \mathbf{V}_B$.

2-8. The acceleration of point A is 18 ft/sec² at 32°, and the acceleration of point B is 12 ft/sec² at 72°. Find $\mathbf{a}_A - \mathbf{a}_B$.

2-9. What was the rate of acceleration of a locomotive which increased its speed from 30 mph to 70 mph while traveling a distance of 3000 ft? How long did it take to increase its speed?

2-10. A plane traveling at 240 mph accelerated at a rate of 12 ft/sec² for a period of 3.7 min. How far did the plane travel during the acceleration interval? What was its final speed?

2-11. A car accelerated from rest to a maximum velocity over a distance of 1500 ft and then decelerated to rest over a distance of 1800 ft. The total time was 99 sec. Determine the maximum velocity reached, the time for the acceleration interval, and the time for the deceleration interval.

2-12. A brake drum decelerated from 740 rpm to 200 rpm in 1.4 sec. Determine the rate of deceleration in rad/sec². Determine also the number of revolutions made by the drum while decelerating.

2-13. How many turns are made by a flywheel which accelerates from an initial speed of 1000 rpm at a rate of 16 rad/sec² for 3.4 sec?

2-14. What is the tangential velocity of a point on the outer surface of a 4 ft diameter rotor in an electric generator when it is turning at 900 rpm?

2-15. What must be the rotational speed in rpm of a cylindrical bar of steel, 3.5 in. D, if the cutting speed is to be 500 ft/min?

2-16. A flywheel, 6.8 ft D, has an instantaneous angular velocity of 600 rpm and an angular acceleration of 42 rad/sec². Determine the normal and tangential components, as well as the total acceleration of a point on the rim.

2-17. A drum 3 ft in diameter accelerated from rest to 300 rpm in 4 sec. What was the total acceleration of a point on the surface of the drum at the end of 2 sec?

2-18. What should be the driving wheel diameter of a locomotive if the speed is to be 75 mph at 500 rpm wheel speed?

2-19. An automobile's tires are 28.4 in. in diameter. What is the car speed in mph when the wheel speed is 600 rpm?

2-20. The wheels in Fig. P2-20 are rolling together without slip. Find the velocity of *A* relative to *B*, of *C* relative to *B*, and of *C* relative to *A*.

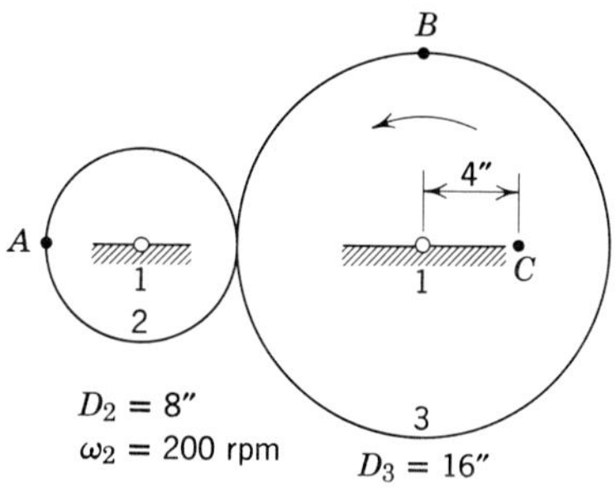

Fig. P2-20

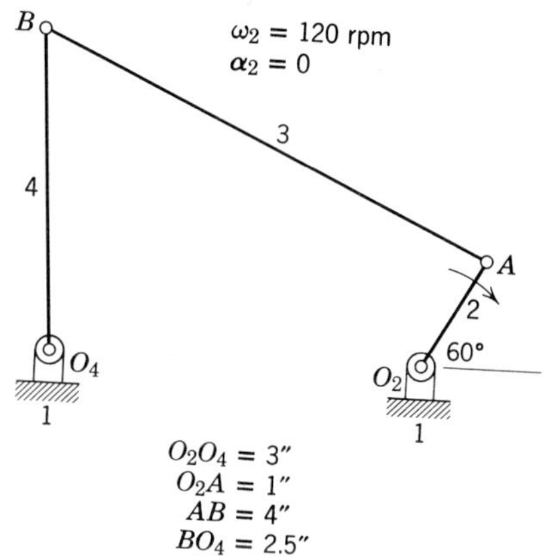

Fig. P2-21

2-21. A four-bar mechanism is shown in Fig. P2-21. Find the velocities of points *A* and *B* and the angular velocities of links 3 and 4.

2-22. What is the actual flight velocity of a plane that heads due North at an air speed of 280 mph if the wind current is moving due East at 35 mph?

2-23. The roller shown in Fig. P2-23 is moving to the right at 15 mph. Find the velocity of *A* relative to *B* and of *C* relative to *A*.

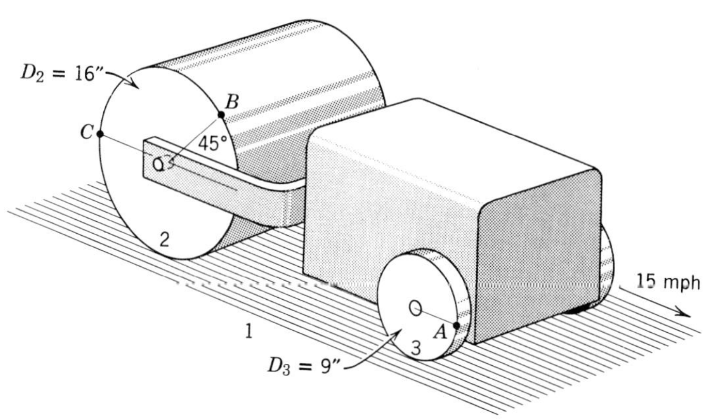

Fig. P2-23

2-24. Plan the layout of the vertical axis of a graph. On it speed is to be represented to the scale 51.5 ft/sec/in. in the range 0 to 200 ft/sec. Decide what scale intervals to use and compute their spacing.

2-25. A Scotch yoke has a 15-in. radius crank turning at 600 rpm clockwise. Determine the period, the maximum yoke displacement, velocity, and acceleration, and the yoke displacement, velocity, and acceleration at the 75° crank position.

2-26. A weight suspended from a tension spring was given an initial deflection and released. It vibrated with simple harmonic motion at 140 cycles per minute with an amplitude of 0.75 in. Determine the maximum velocity and acceleration of the weight.

2-27. Below is a table showing the slider displacement from top dead center of an in-line slider-crank mechanism. The crank radius is 5 in. and the connecting rod length 17.5 in. The crank turns counterclockwise at 600 rpm. Calculate the time corresponding to each crank position, taking $t = 0$ at $\theta = 0$, and plot the displacement-time curve. Then, by graphical differentiation, plot the velocity-time curve.

Crank Angle, degrees	Displacement, inches
0 and 360	0.000
15 and 345	0.218
30 and 330	0.849
45 and 315	1.825
60 and 300	3.044
75 and 285	4.386
90 and 270	5.730
105 and 255	6.974
120 and 240	8.044
135 and 225	8.896
150 and 210	9.510
165 and 195	9.878
180	10.000

3 Velocity Analysis

3-1. Introduction

One of the most important tasks in the analysis or design of a mechanism is the determination of the velocities of key points and links. It is the purpose of this chapter to present two methods which have proved useful and efficient in making velocity analyses.

There are a number of motives for making a velocity analysis. In many mechanisms the velocity of some member is itself the most important feature in the operation of the mechanism. Close control of the shuttle velocity in a textile machine may be the chief goal, or perhaps it is the shaft speed in a timing device that is crucial. In a machine tool it will be important to determine the cutting speed of the tool at various positions of the driving crank.

In machines such as pile drivers and forging presses, the kinetic energy of certain parts is of prime interest. And since velocity determines kinetic energy, a knowledge of the velocities of these parts is necessary to the design.

A second motive for making a velocity analysis is as a necessary first step in an acceleration analysis. Acceleration analysis and the motives for making it form the subject matter of Chapter 4.

In each of the methods of velocity analysis presented in this chapter, the analysis being carried out will occur at a particular phase of the mechanism. It is assumed that for each phase of the mechanism, its geometry and the velocity of some point or link are known. A complete velocity analysis is made by a repetition of the process for each phase in which the designer is interested.

3-2. Instant Centers

Definition. A body m having plane motion relative to a reference body n (not necessarily fixed) is shown in Fig. 3-1. The velocities of two points on m, A and B, relative to n are known. Applying Eq. 2-25, vector equations can be written relating the velocity of any third point, P, to those of A and B.

$$\mathbf{V}_P = \mathbf{V}_A + \mathbf{V}_{P/A} \tag{3-1}$$

$$\mathbf{V}_P = \mathbf{V}_B + \mathbf{V}_{P/B} \tag{3-2}$$

Let P be the point at the intersection of lines from A and B, perpendicular, respectively, to $\mathbf{V}_A$ and $\mathbf{V}_B$. Then, since $\mathbf{V}_A$ and $\mathbf{V}_{P/A}$ are each perpendicular to line PA, $\mathbf{V}_P$ must also be. However, since $\mathbf{V}_B$ and $\mathbf{V}_{P/B}$ are each perpendicular to line PB, $\mathbf{V}_P$ must be also. This is a contradiction: $\mathbf{V}_P$ cannot, in general, be perpendicular to both lines PA and PB. The apparent contradiction is resolved if

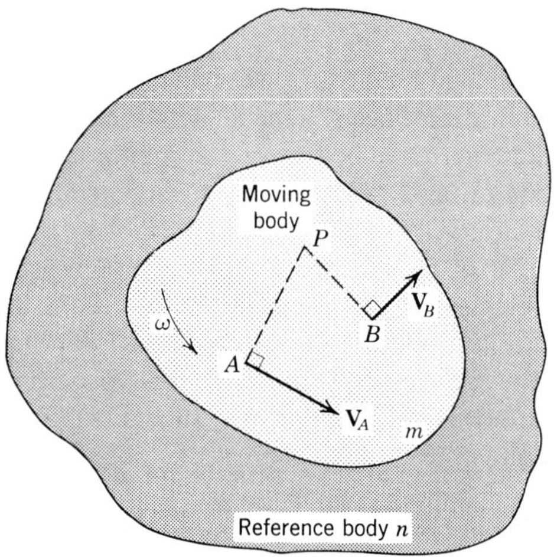

Fig. 3-1. Instant center.

$$\mathbf{V}_P = 0 \tag{3-3}$$

Point P is called an *instant center* (or *instantaneous center*, *rotopole*, *centro*). One of its properties is now clear. If two bodies are having relative plane motion, the instant center is the point at which they are relatively at rest.

Since $\mathbf{V}_P = 0$, then from Eqs. 3-1 and 3-2,

$$\mathbf{V}_A = -\mathbf{V}_{P/A} = \mathbf{V}_{A/P}$$

$$\mathbf{V}_B = -\mathbf{V}_{P/B} = \mathbf{V}_{B/P}$$

But from Eq. 2-26,

$$V_{A/P} = \omega(AP)$$

and $$V_{B/P} = \omega(BP)$$

Therefore, $$V_A = \omega(AP) \tag{3-4}$$

$$V_B = \omega(BP) \tag{3-5}$$

and $$\frac{V_B}{V_A} = \frac{BP}{AP} \tag{3-6}$$

Equations 3-4 and 3-5 reveal the second basic property of an instant center. At the instant, the two bodies may be regarded as having relative rotation about the instant center. In summary, the basic properties of an *instant center* are:

1. It is the point at which the two bodies having plane relative motion are relatively at rest.
2. It is the point about which the bodies are rotating relative to one another at the instant.

The first of the above properties has an immediate implication which will be of value in the velocity analyses of mechanisms: the velocity of the instant center relative to any third body will be the same whether regarded as on the first or on the second body.

The potential usefulness of instant centers is indicated by Eq. 3-6. Stating this equation as a rule:

> The magnitudes of the velocities of points on a rigid body having plane motion vary directly as the distances from the points to the instant center.

Thus if the instant center and the velocity of one point on a link are known, the velocity of any other point on the same link can be found.

Number of instant centers. For each pair of bodies having relative plane motion there is an instant center. Therefore in a mechanism there are as many instant centers as there are possible pairs of links. For any mechanism the equation below gives the number of instant centers:

$$N = \frac{n(n-1)}{2} \tag{3-7}$$

where N = the number of instant centers
n = the number of links

Example 3-1. How many instant centers are there for a mechanism having 8 links?

Solution. Since $n = 8$, applying Eq. 3-7,

$$N = \frac{8(8 - 1)}{2}$$
$$= 28$$

Notation. Figure 3-2 indicates a number of the ways in which instant centers are commonly designated. The notations of Figs. 3.2*c* and 3.2*f* are the more economical and will usually be employed in this text.

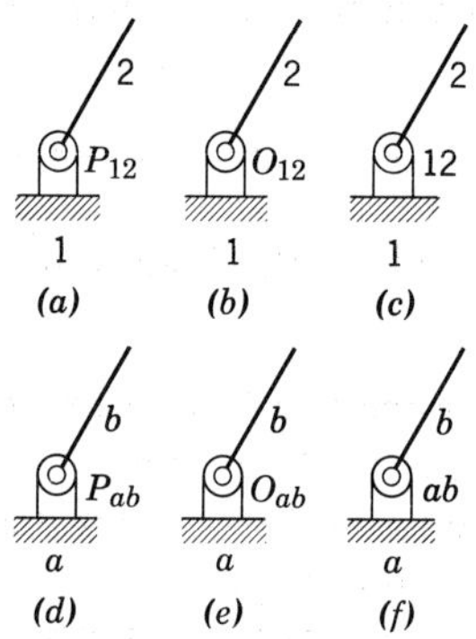

Fig. 3-2. Instant center notation.

3-3. The Aronhold-Kennedy Theorem

This theorem, which is also known as the *Kennedy Theorem* or the *Kennedy Rule*, states:

If three bodies have relative plane motion, they have three instant centers which lie on the same straight line.

Proof

1. That three bodies having relative plane motion have three instant centers follows from Eq. 3-7. Setting $n = 3$,

$$N = \frac{3(3 - 1)}{2}$$
$$= 3$$

It remains to prove that they lie on the same straight line.

2. In Fig. 3-3 there are three bodies having relative plane motion. The locations of instant centers 12 and 13 are known. It is to be proved that 23 lies on line 12-13. This will be done by showing that to assume it does not lie on the line leads to a contradiction.

3. A position for 23 has been assumed in Fig. 3-3 off line 12-13. Now, regarding point 23 as on body 2,

$$\mathbf{V}_{23} \perp 12\text{-}23$$

Regarding 23 as on body 3,

$$\mathbf{V}_{23} \perp 13\text{-}23$$

4. But it follows from the properties of an instant center that the velocity of the instant center for two bodies relative to a third is the same whether

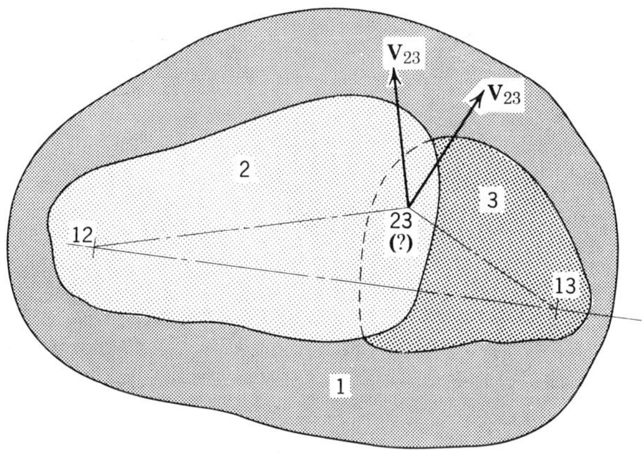

Fig. 3-3. Proof of the Aronhold-Kennedy Theorem.

regarded as a point on the first or on the second. The velocity of 23 cannot be perpendicular to both 12-23 and 13-23 unless point 23 lies on line 12-13.

3-4. Locating Instant Centers

Rules. The following rules are used in locating instant centers:

1. If two links are connected by a pivot joint, the center of the pivot is their instant center.
2. If two links have pure rolling contact, the point of contact is their instant center.
3. If two links have sliding contact, their instant center is along the common normal at the point of contact.
4. The three instant centers for any three links having relative plane motion lie on the same straight line (Aronhold-Kennedy Theorem).

These rules permit finding some instant centers by inspection, the *primary* instant centers. In any mechanism the primary instant centers must be located first, the rest being found by constructions.

Turning joints. The most obvious of the instant centers located by inspection are those at the centers of turning joints, as in Fig. 3-4*a*. If two links are pinned together, each will rotate relative to the other about the center of the joint. In Fig. 3-4*a* the joint is a permanent one; hence the instant center is called *permanent*. If, in addition, one of the links is fixed, the instant center is *fixed*.

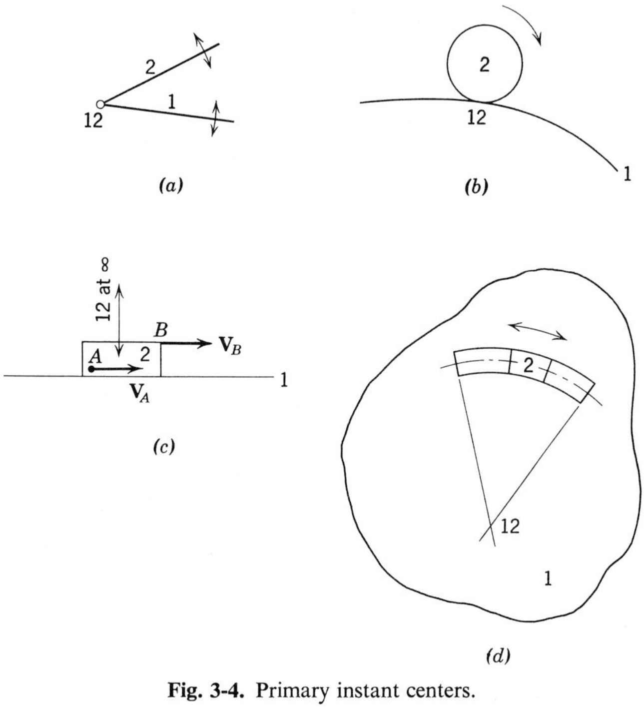

Fig. 3-4. Primary instant centers.

Rolling contact. If two links are having rolling contact, this means by definition that there is no sliding between the surfaces. Therefore the points of the two links which are in contact must have the same velocity. They may be regarded, when convenient, as a single point of contact common to both bodies. The instant center is this point of contact, as in Fig. 3-4*b*.

Sliding contact. Some of the instant centers for pairs of links having sliding contact can be located by inspection. Figure 3-4*c* shows a slider moving on a flat surface. Each point on the slider has the same velocity: it is a case of rectilinear translation. If lines are drawn from any two points on the slider, *A* and *B*, perpendicular to $\mathbf{V}_A$ and $\mathbf{V}_B$, instead of intersecting at the instant center, they are parallel. In this case it is convenient to say that the instant center 12 is "at infinity," in a direction perpendicular to the path of motion of the slider.

If a slider moves in a circular slot, as shown in Fig. 3-4*d*, the instant center is at the center of the circular slot. The motion of the slider is kinematically the same as that of a link rotating about a turning joint at 12.

The general case of sliding contact is illustrated in Fig. 3-5. The instant

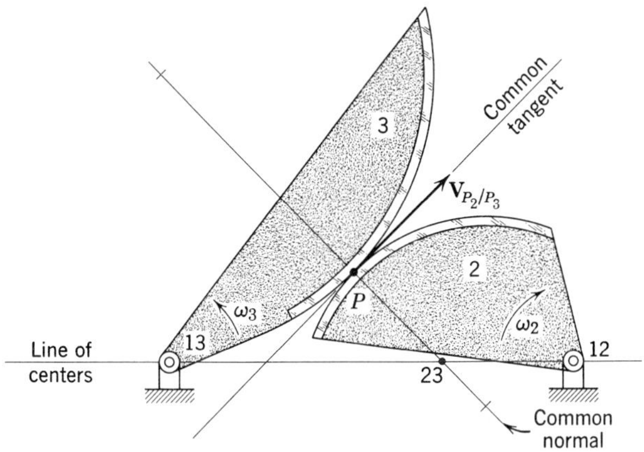

Fig. 3-5. Instant centers for direct contact mechanism.

center must be on the common normal to the surfaces at the point of contact.

Proof

1. The points of contact are P_2 on 2 and P_3 on 3. Their relative velocity must be along the common tangent. If it were not, it could be resolved into components along the common tangent and the common normal. But if there were a component along the common normal, this would imply that the links were either separating or overlapping. Hence the relative velocity must be along the common tangent. $\mathbf{V}_{P_2/P_3}$ has been shown in the figure along the common tangent.

2. But if the velocity of P_2 relative to P_3 is along the common tangent, then the instant center for the two bodies must be along the common normal. When a body is rotating about a point, the velocity of any point on the body is perpendicular to the radius to the point.

To locate instant center 23 completely, the Aronhold-Kennedy must also be employed. The instant centers for links 1, 2, and 3 must lie on the same line. Thus instant center 23 must lie on line 12-13. But it also lies on the common normal. Therefore 23 lies at the intersection of the line of centers, 12-13, and the common normal.

Instant center polygons. When a mechanism has four or more links, some convenient means is needed for keeping track of those instant centers which have already been found and for developing a strategy to find any instant center not yet located. The instant center polygon is one such means.

It consists in a regular polygon having as many sides as the mechanism has links. In such a figure there will be one side or diagonal for each pair

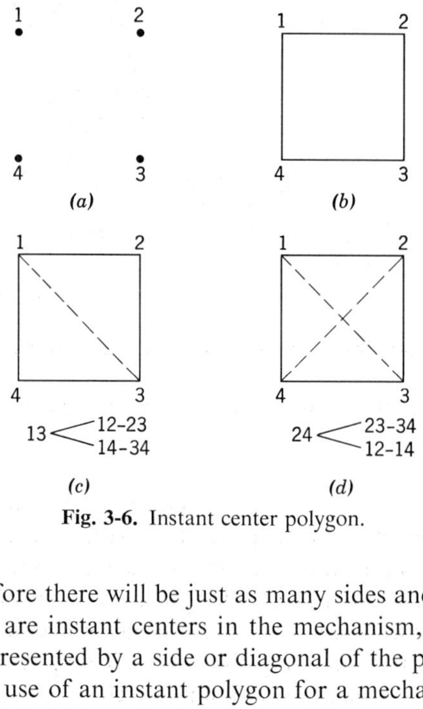

Fig. 3-6. Instant center polygon.

of points. Therefore there will be just as many sides and diagonals in the polygon as there are instant centers in the mechanism, and each instant center will be represented by a side or diagonal of the polygon.

In Fig. 3-6 the use of an instant polygon for a mechanism having four links is illustrated. Initially only the corners of the square are drawn. as in Fig. 3-6*a*, and numbered or lettered using the names of the links. The numbering can be done either clockwise or counterclockwise.

Next the mechanism layout is inspected, and all those instant centers which can be found by inspection—the primary centers—are found and labeled. In the polygon a solid line is drawn in for each instant center thus located. For example, instant centers 12, 23, 34, and 14 were found by inspection, and thus solid lines have been drawn in Fig. 3-4*b*.

The remaining instant centers, 13 and 24, can be found by an application of the Aronhold-Kennedy Theorem. Line 13 in the polygon (Fig. 3-6*c*) is drawn as a dashed line because it is located by construction. Line 13 "closes" triangle 123, indicating: (*a*) that instant center 13 lies on line 12-23; (*b*) that instant centers 12 and 23 have been located previously, and line 12-23 can therefore actually be drawn.

Line 13 also closes triangle 134, indicating that 13 lies on line 14-34, and that this line can be drawn. The schema

$$13 \left\langle \begin{array}{l} 12\text{-}23 \\ \\ 14\text{-}34 \end{array} \right.$$

is a short-hand way of recording the fact that instant center 13 is found at the intersection of lines 12-23 and 14-34.

The strategy for locating instant center 24 is indicated in Fig. 3-6*d*. Since 24 closes triangles 234 and 241, 24 lies at the intersection of lines 12-14 and 23-34.

3-5. Examples of Instant Center Location

In this article, through examples, the application of the rules for the location of instant centers will be demonstrated.

Example 3-2. Find the instant centers for the four-bar mechanism in Fig. 3-7*a*.

Solution

1. The number of instant centers is found by using Eq. 3-7.

$$N = \frac{4(4-1)}{2}$$
$$= 6$$

Since there are four links, the instant center polygon will be a square. The corners of the square have been indicated and labeled in Fig. 3-7*a*.

2. Next the primary instant centers are located. There are four primary centers, the four pin joints 12, 23, 34, and 14. They have been labeled in Fig. 3-7*b*, and solid lines have been drawn in the polygon.

3. In the polygon, 13 closes triangles 123 and 143. Therefore 13 lies at the intersection of lines 12-23 and 14-34. It has been so located in Fig. 3-7*c*.

4. Line 24 in the polygon closes triangles 214 and 234. Therefore 24 lies at the intersection of lines 12-14 and 23-34. It has been located in Fig. 3-7*d*.

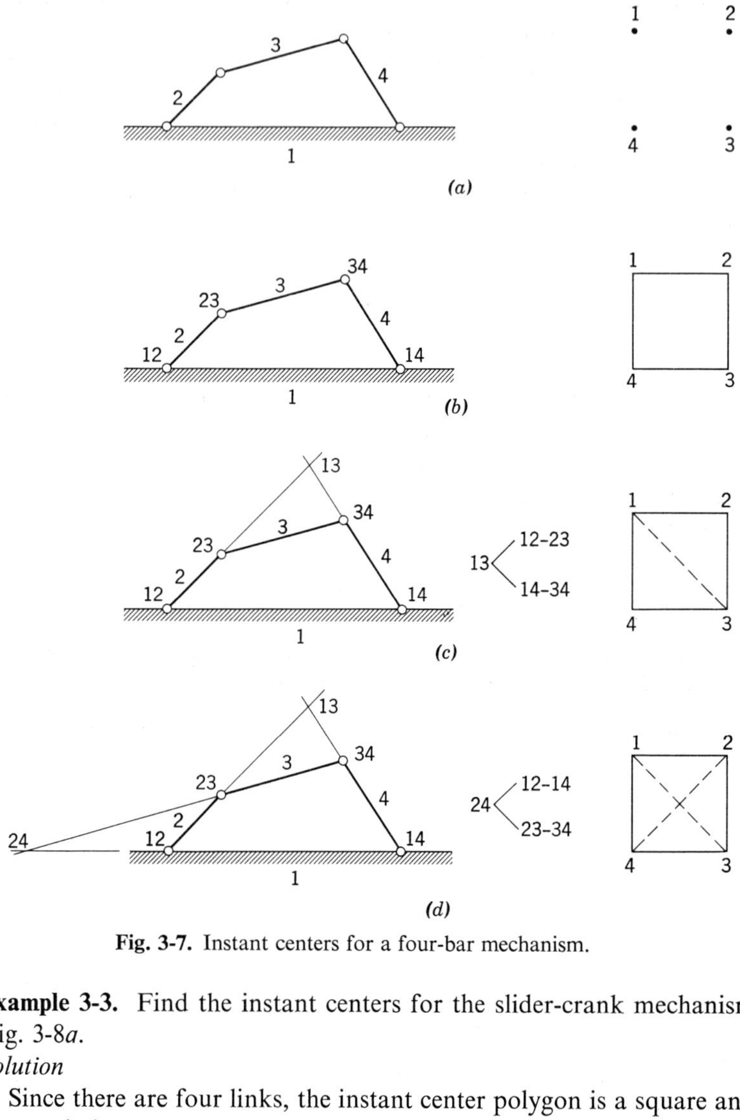

Fig. 3-7. Instant centers for a four-bar mechanism.

Example 3-3. Find the instant centers for the slider-crank mechanism in Fig. 3-8*a*.

Solution

1. Since there are four links, the instant center polygon is a square and there are six instant centers.

$$N = \frac{4(4-1)}{2}$$
$$= 6$$

2. The primary instant centers have been found in Fig. 3-8*b*. There are three pin joints and one sliding pair. The instant center 14 is "at infinity"

in a direction perpendicular to the slider's motion. Instant centers 12, 23, and 34 are the centers of pin joints.

3. Instant center 24 is at the intersection of lines 12-14 and 23-34 since line 24 in the polygon closes triangles 124 and 234. Line 12-14 is drawn

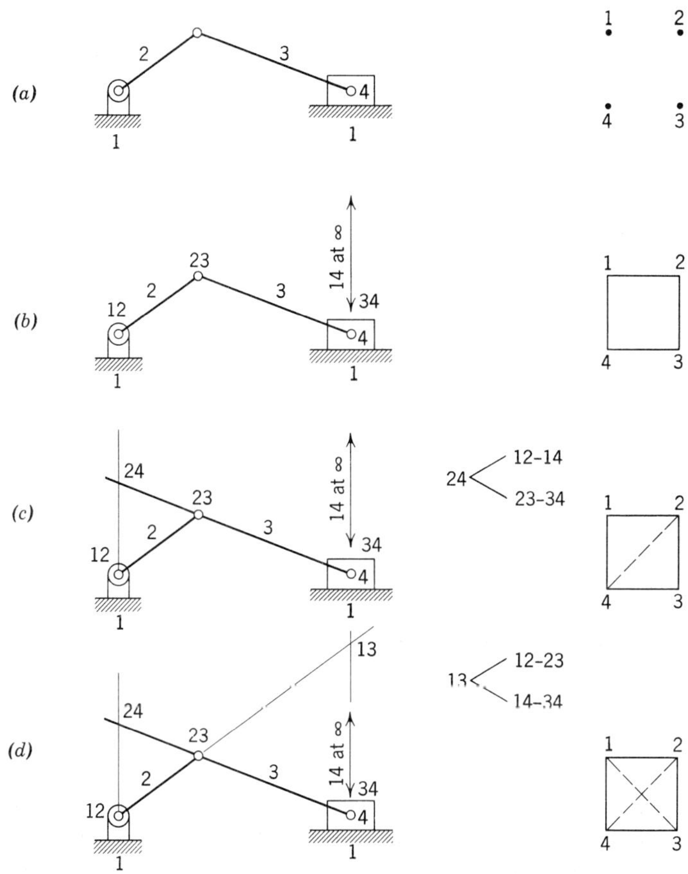

Fig. 3-8. Instant centers for a slider-crank mechanism.

from 12 and perpendicular to the direction of motion of the slider, since 14 is at infinity. Line 23-34 extended intersects line 12-14 at 24.

4. Line 13 closes triangles 143 and 123; so 13 lies at the intersection of 14-34 and 12-23. Line 14-34 is drawn from 34 and perpendicular to the direction of motion of the slider, since 14 is at infinity in a direction perpendicular to the slider's motion direction. Where line 12-23 produced intersects line 14-34 is instant center 13.

Example 3-4. Find the instant centers for the three-link cam mechanism of Fig. 3-9.

Solution

1. Because there are only three links, there are only three instant centers and an instant center polygon is unnecessary.

2. Two instant centers, 12 and 13, are located by inspection alone. 12 is where the cam pivots relative to the frame. Since the follower, link 3, has a sliding motion relative to the frame which is rectilinear, instant center 13 is at infinity in a direction perpendicular to the direction of motion of the follower.

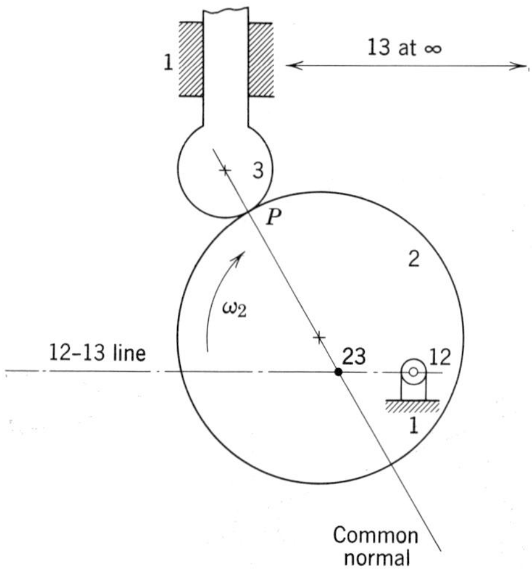

Fig. 3-9. Instant centers for a three-link cam mechanism.

3. It remains to find 23. Since links 2 and 3 have sliding contact, 23 must lie on the common normal to the surfaces. But it must also lie on line 12-13. Therefore it lies at their intersection. The common normal has been drawn through the centers of curvature of the two contacting surfaces, which in this case are each circular. Line 12-13 has been drawn from 12 and in a direction perpendicular to the direction of motion of the follower. The intersection of these two lines is instant center 23.

Example 3-5. Find the instant centers for the four-link cam mechanism shown in Fig. 3-10*a*.

Solution

1. There are four links. The polygon will be a square and there will be six instant centers.

2. Four of the six instant centers can be located by inspection. The cam to frame pivot is 12, and the roller pivot is 34. There will be pure rolling contact between the roller and the cam; so instant center 23 is at their

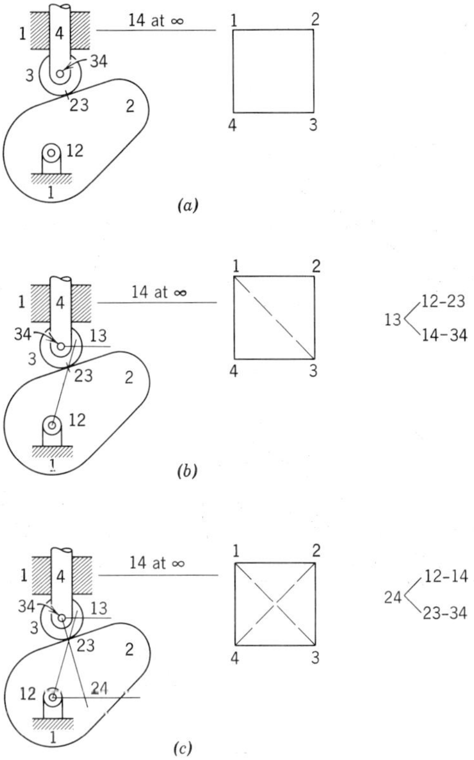

Fig. 3-10. Instant centers for a four-link cam mechanism.

point of contact. And instant center 14 is at infinity in a direction perpendicular to the direction of motion of the follower body relative to the frame.

3. Line 13 closes triangles 123 and 143. Therefore 13 lies at the intersection of 14-34 and 12-23.

4. Similarly, 24 lies at the intersection of lines 12-14 and 23-34.

Example 3-6. Find the instant centers for the five-link mechanism of Fig. 3-11*a*.

Solution. The primary instant centers have been located in Fig. 3-11*b*, and the remaining centers found by construction in Fig. 3-11*c*. The reader should follow through these constructions for himself.

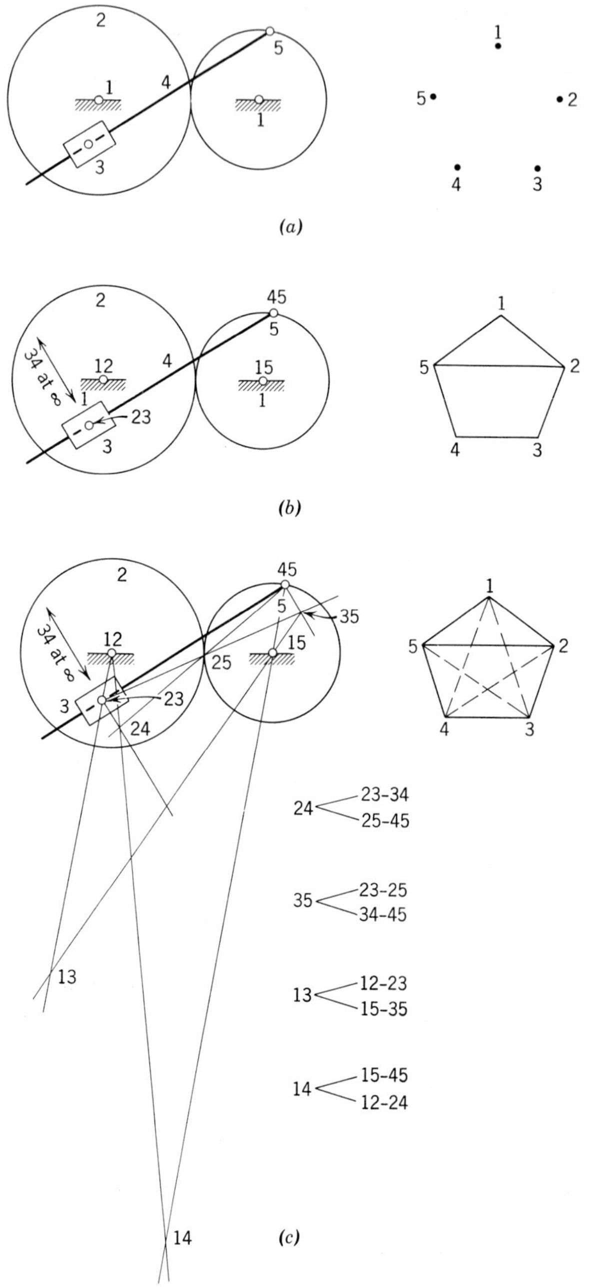

Fig. 3-11. Instant centers in a five-link mechanism.

3-6. Inaccessible Instant Centers

Sometimes, when one is locating an instant center by construction, it falls at the intersection of two lines, which, being nearly parallel, intersect well off the paper. Yet it may be necessary to draw the line determined by this inaccessible instant center and some other instant center. A technique for accomplishing this is illustrated by the following example.

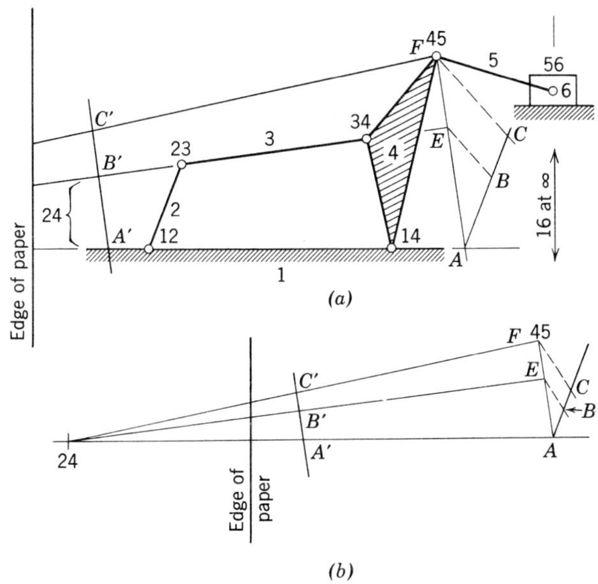

Fig. 3-12. Drawing to an instant center off the paper.

Example 3-7. In Fig. 3-12*a* a six-link mechanism is shown. Instant center 24 is at the intersection of lines 23-34 and 12-14 and is inaccessible. It is required to draw the line determined by 45 and 24.

Solution. The solution is shown in Fig. 3-12*a*.

1. From 45 draw any line intersecting line 12-14 at A and 23-34 at E.
2. At a convenient position near the edge of the paper, draw a line parallel to 45-A, intersecting 12-14 at A' and 23-34 at B'.
3. From A draw any line, and on it locate B so that $AB = A'B'$.
4. From 45 draw a line parallel to EB, intersecting line AB at C.
5. On line $A'B'$ locate C' so that $A'C' = AC$.
6. The line 45-C' is the required one.

Proof. The construction has been shown with the mechanism details omitted in Fig. 3-12*b*. For convenience, point 45 has been labeled also as F.

1. By construction, lines FC and EB are parallel. Therefore triangles ABE and ACF are similar, and

$$\frac{AC}{AB} = \frac{AF}{AE}$$

2. But, by construction, $AB = A'B'$ and $AC = A'C'$. Therefore,

$$\frac{AF}{AE} = \frac{A'C'}{A'B'}$$

3. If C' is on line 24-45, then triangles $24A'C'$ and $24AF$ are similar, as are triangles $24AE$ and $24A'B'$. Therefore,

$$\frac{AF}{24A} = \frac{A'C'}{24A'}$$

or

$$\frac{AF}{A'C'} = \frac{24A}{24A'}$$

and

$$\frac{AE}{24A} = \frac{A'B'}{24A'}$$

or

$$\frac{AE}{A'B'} = \frac{24A}{24A'}$$

4. But then

$$\frac{AF}{A'C'} = \frac{AE}{A'B'}$$

or

$$\frac{AF}{AE} = \frac{A'C'}{A'B'}$$

Since this proportionality checks with that of Step 2, the assumption that C' is on line 45-24 is justified.

3-7. Velocity Analysis by Instant Centers

Basic construction. It has been shown that the velocities of points on a link have magnitudes directly proportional to the distances from the points to the instant center.

In Fig. 3-13*a*, body 2 has plane motion relative to body 1, and the instant center is 12. The velocity of point A is known, and that of point B is required. The procedure is based upon the use of similar triangles and the fact that points the same distance from the instant center will have velocities of equal magnitudes. Two slightly different constructions are shown in Figs. 3-13*b* and 3-13*c*. The choice between them is a matter of convenience.

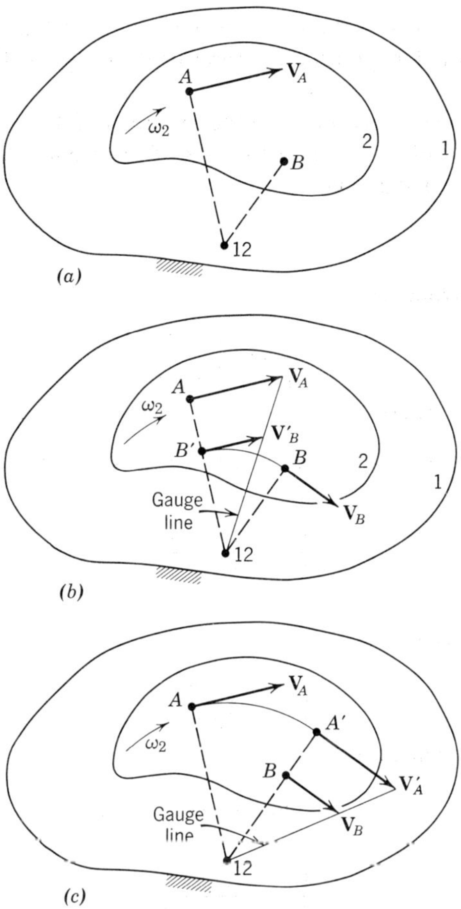

Fig. 3-13. Velocity analysis by instant centers.

In Fig. 3-13*b*:

1. Revolve point B to a position B' on radial line 12-A.
2. Draw a gauge line from 12 to the terminus of $\mathbf{V}_A$.
3. From B' draw a vector perpendicular to 12-B' and terminating at the gauge line. This is $\mathbf{V}_{B'}$.
4. Since points B and B' are the same distance from instant center 12, their velocities have equal magnitudes. Therefore draw $\mathbf{V}_B$ at B, perpendicular to 12-B, and so that $V_B = V_{B'}$.

In Fig. 3-13*c*:

1. Revolve A to position A' on radial line 12-B.

2. Since A and A' are the same distance from instant center 12, their velocities have equal magnitudes. Therefore draw $\mathbf{V}_{A'}$ at A', perpendicular to 12-A', and such that $V_{A'} = V_A$.

3. Draw a gauge line from 12 to the terminus of $\mathbf{V}_{A'}$.

4. The velocity of B is the vector from B, perpendicular to 12-B, and terminating at the gauge line.

Proofs

In Fig. 3-13*b*:

1. By similar triangles,

$$\frac{V_{B'}}{V_A} = \frac{12 - B'}{12 - A}$$

2. But, by construction,

$$V_B = V_{B'} \quad \text{and} \quad 12 - B = 12 - B'$$

3. Therefore,

$$\frac{V_B}{V_A} = \frac{12 - B}{12 - A}$$

In Fig. 3-13*c*:

1. By similar triangles,

$$\frac{V_B}{V_{A'}} = \frac{12 - B}{12 - A'}$$

2. But, by construction,

$$V_A = V_{A'} \quad \text{and} \quad 12 - A = 12 - A'$$

3. Therefore,

$$\frac{V_B}{V_A} = \frac{12 - B}{12 - A}$$

Since the proportionality in Step 3 of each proof states correctly that the magnitudes of the velocities of two points on the same link are directly proportional to the distances from the points to the instant center, each construction is valid.

Direct method. The instant center technique for velocity analysis permits only the determination of the velocities of other points on the *same link* if the instant center and the velocity of one point on the link are given. How then can the velocity of a point on another link be found? The answer is that first the velocity of a point common to the two links must be found. The most convenient common point to use, and frequently the only one, is the instant center for the two links.

For example, suppose that the velocity of point P on link 3 is known, and it is required to find the velocity of point Q on link 7. The solution has two parts:

1. First the velocity of instant center 37 is found from that of P, where P and 37 are points on link 3.
2. Second the velocity of point Q is found from that of point 37, where 37 and Q are points on link 7.

The above two-part strategy is justified by recalling that an instant center is a point common to the two links, and its velocity is the same whether regarded as on the first or on the second of these two links.

The techniques of velocity analysis using instant centers will be illustrated by examples in the next article.

3-8. Examples of Velocity Analysis by Instant Centers

In these examples, references to actual numerical values have been omitted for simplicity. In addition, the constructions for instant centers have usually not been shown; however, it is assumed, of course, that the instant centers had previously been located and are thus available when needed.

Example 3-8. The velocity of point A in a four-bar mechanism is shown in Fig. 3-14*a*. The velocities of points B and C are to be found.

Solution. Since points B and C are both on link 4, the strategy followed will be to determine the velocity of instant center 24 first. Then the velocities of points B and C can be found. Because A is also on link 3, the velocity of 34 could have been found instead.

1. Revolve A about 12 to position A' on radial line 12-24. The construction will be based at 12, for points 24 and A are both on link 2, and link 2 turns relative to ground at 12.
2. Lay out $\mathbf{V}_{A'}$ at A'. The magnitude of $\mathbf{V}_{A'}$ is the same as that of $\mathbf{V}_A$.
3. Draw a gauge line from 12 through the terminus of $\mathbf{V}_{A'}$.
4. $\mathbf{V}_{24}$ is the vector at 24, perpendicular to 12-24, and terminating at the gauge line.
5. Since points 24, B, and C are all on link 4, the next construction will be based at 14, where link 4 turns relative to the frame. Therefore draw a gauge line from 14 to the terminus of $\mathbf{V}_{24}$.
6. Revolve B and C about 14 to positions B' and C' on line 14-24.
7. At B' draw $V_{B'}$, perpendicular to the 14-12 line, and terminating at the gauge line. Similarly draw $\mathbf{V}_{B'}$.

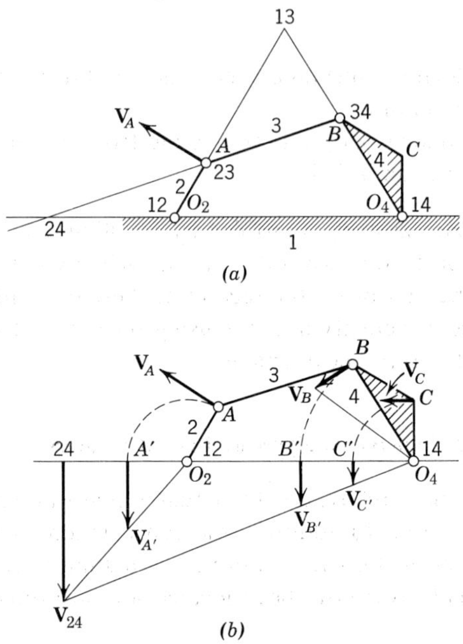

Fig. 3-14. Instant center velocity analysis of four-bar linkage.

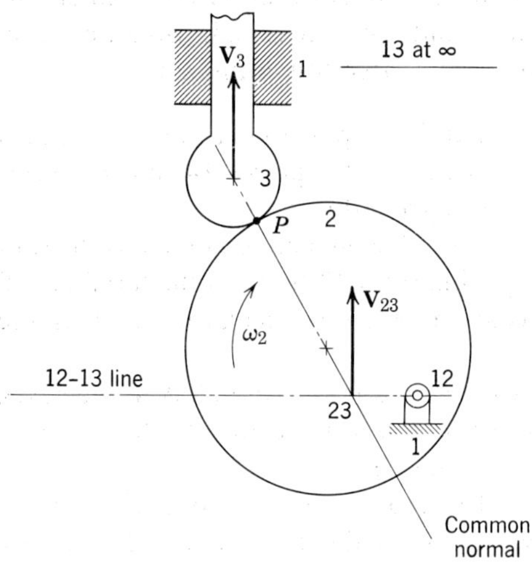

Fig. 3-15. Instant center velocity analysis of simple cam mechanism.

8. Then $\mathbf{V}_B$ and $\mathbf{V}_C$ can be drawn. $\mathbf{V}_B$ is perpendicular to line 14-B and $\mathbf{V}_C$ to line C-14. The magnitudes of $\mathbf{V}_B$ and $\mathbf{V}_C$ are the same as those of $\mathbf{V}_{B'}$ and $\mathbf{V}_{C'}$. The senses of $\mathbf{V}_B$ and $\mathbf{V}_C$ must be consistent with that of $\mathbf{V}_{24}$.

Example 3-9. A three-link cam mechanism is shown in Fig. 3-15. The angular velocity of the cam is given, and it is required to find the velocity of the follower.

Solution

1. Since link 3 is a slider having rectilinear translation, each point on it has the same velocity. But instant center 23 can be considered as a point on the slider. Therefore the velocity of instant center 23 is the same as that of the slider.

2. Instant center 23 can also be considered as a point on the cam, link 2. Therefore,

$$V_{23} = \omega_2(12\text{-}23)$$

The sense of $\mathbf{V}_{23}$ must be upward, since the cam is turning clockwise.

3. $\mathbf{V}_3 = \mathbf{V}_{23}$.

Example 3-10. A six-link mechanism is shown in Fig. 3-16a. The velocity of A is given; it is required to find the velocities of the slider and point E.

Solution

To find the slider velocity:

1. Revolve A about 12 to position A' on line 12-26. This is done in order to find the velocity of point 26, a point on the slider.

2. At A' lay out $\mathbf{V}_{A'}$.

3. Draw a gauge line from 12 to the terminus of $\mathbf{V}_{A'}$.

4. The vector at 26, perpendicular to line 12-26, and terminating at the gauge line is $\mathbf{V}_{26}$.

5. Since 26 can be regarded as a point on the slider, and since the slider has rectilinear translation, $\mathbf{V}_6 = \mathbf{V}_{26}$.

To find the velocity of point E:

1. Revolve 24 about 12 to a position 24′ on line 12-26.

2. The vector at 24′, perpendicular to the line 12-26, and terminating at the gauge line is $\mathbf{V}_{24'}$.

3. Revolve 24 about 14 to a position 24″ on line 14-E.

4. Lay out the velocity of 24″ having the same magnitude as that of 24 and 24′.

5. Through 14 and the terminus of $\mathbf{V}_{24''}$ draw a gauge line.

6. The vector at E, perpendicular to 14-E, and terminating at the gauge line is $\mathbf{V}_E$.

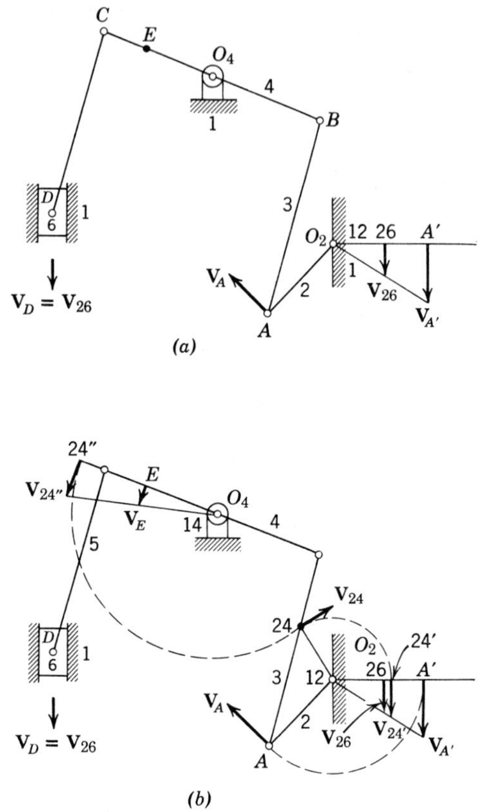

Fig. 3-16. Instant center velocity analysis of a six-link mechanism.

3-9. Parallel Line Construction

When an instant center at which a similar triangle construction is to be based in a velocity analysis is inaccessible, an alternative method is needed.

Figure 3-17 shows a four-bar mechanism. The velocity of A is given and the velocity of B is required. The usual procedure, since points A and B are both on link 3, would be to make a similar triangle construction based at instant center 13. But 13 is not available; the following construction can be used instead:

1. Rotate $\mathbf{V}_A$ about its origin 90° to position AA' on radial line 13-A.
2. From A' draw a line parallel to the line joining points A and B to intersect radial line 13-B at B'. BB' represents the magnitude of the velocity of B to the same scale that AA' represents that of A.

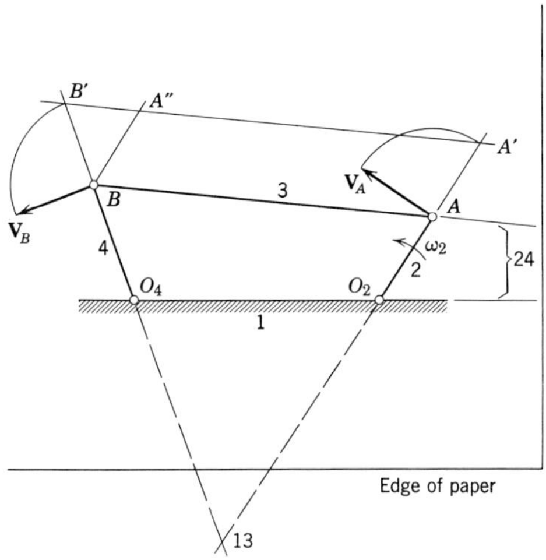

Fig. 3-17. Parallel line construction.

Proof

1. Draw a line from B parallel to AA', intersecting $A'B'$ at A''.
2. Then since triangles $BB'A''$ and 13-BA are similar,

$$\frac{B'B}{BA''} = \frac{B\text{-}13}{A\text{-}13}$$

3. But

$$\frac{V_B}{V_A} = \frac{B\text{-}13}{A\text{-}13}$$

4. Therefore,

$$\frac{V_B}{V_A} = \frac{BB'}{BA''}$$

5. Since $BA'' = AA'$,

$$\frac{V_B}{V_A} = \frac{BB'}{AA'}$$

6. But since $V_A = k_V \cdot AA'$,

$$\frac{V_B}{k_v \cdot AA'} = \frac{BB'}{AA'}$$

or

$$V_B = k_v \cdot BB'$$

That is, BB' represents V_B to the scale that AA' represents V_A.

3-10. Angular Velocities

Rules. Two bodies having plane motion relative to a third are shown in Fig. 3-18*a*. The three instant centers are known. They are, of course,

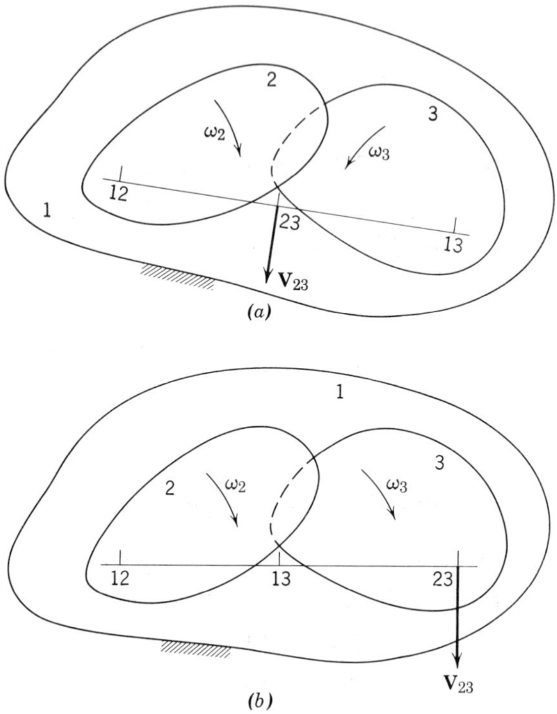

Fig. 3-18. Angular velocities of links.

collinear. Since instant center 23 has the same velocity relative to body 1 whether regarded as a point on 2 or on 3,

$$V_{23} = (13\text{-}23)\omega_3$$

and

$$V_{23} = (12\text{-}23)\omega_2$$

Therefore,

$$\frac{\omega_3}{\omega_2} = \frac{12\text{-}23}{13\text{-}23}$$

If bodies 1, 2, and 3 are links of a plane mechanism, where link 1 is the frame, the above equation can be restated as a rule:

> The angular speeds of two links in a plane mechanism vary inversely as the distances from their centers of rotation relative to the frame to their common instant center.

Using this rule, if the instant centers and the angular speed of one link in a plane mechanism are known, the angular speed of any other link can be found.

A second rule is needed which will relate the directions of rotation of the links. An inspection and comparison of the two parts of Fig. 3-18 will provide the rule. In Fig. 3-18*a*, if link 2 turns clockwise, the velocity of 23 must be downward. The reason is that, since body 2 turns about 12 and 23 is to the right of 12, only a downward direction for $\mathbf{V}_{23}$ could be consistent with the clockwise rotation of the body. But if the velocity of 23 is down, this in turn implies that link 3 must be turning counterclockwise. In Fig. 3-18*b* a clockwise rotation of link 2 implies a downward direction for the velocity of 23, which in turn implies that link 3 must be turning in a clockwise direction. In Fig. 3-18*a* the common instant center lies between the ground centers, while in Fig. 3-18*b* it is outside this line segment. These observations can be generalized in the following rule:

> If the common instant center for two links lies on the line segment joining the instant centers of the two links relative to the frame, the links turn in opposite directions. If the common instant center lies outside this segment, they turn in the same direction.

Graphical construction. The construction is illustrated in both Fig. 3-19*a* and 3-19*b*. In the first, the common instant center lies between the ground centers; in the second it is outside. The construction is the same for each case. In each, the instant centers and the angular velocity of one link, link 2, are given. It is required to find the angular velocity of link 3. Link 1 is the frame. The procedure is:

1. At 13 lay out a line 13-A, perpendicular to the line of centers, and representing ω_2 to some convenient scale.
2. Draw a gauge line through *A* and 23.
3. At 12 erect a line perpendicular to the line of centers and terminating at *B* on the gauge line. 12-*B* represents ω_3 to the same scale that 13-*A* represents ω_2.
4. Examine the layout to determine the direction of the link's movement. In Fig. 3-19*a*, link 3 must be turning in a direction opposite that of link 2, for 23 falls between 12 and 13. In Fig. 3-19*b* the links turn in the same direction because 23 lies outside the line segment 12-13.

At first glance it may seem strange that the speed of link 2 was drawn at 13 and that of link 3 at 12. This is because the speeds vary *inversely* as the distances to the common instant center.

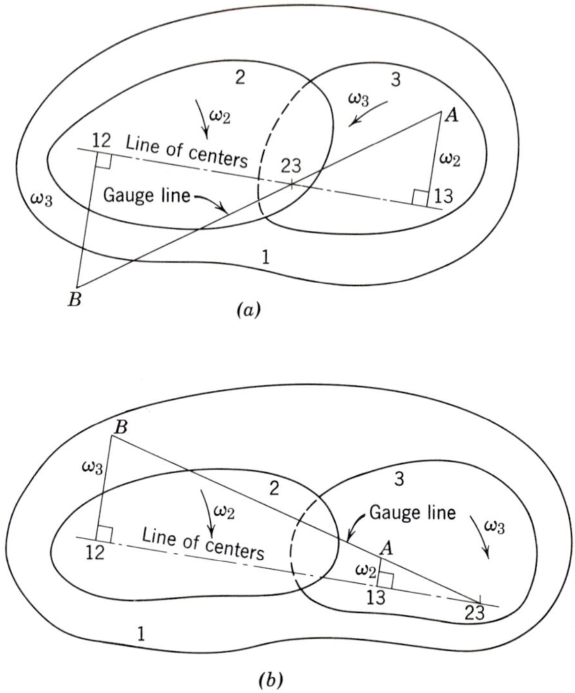

Fig. 3-19. Graphical determination of angular speeds.

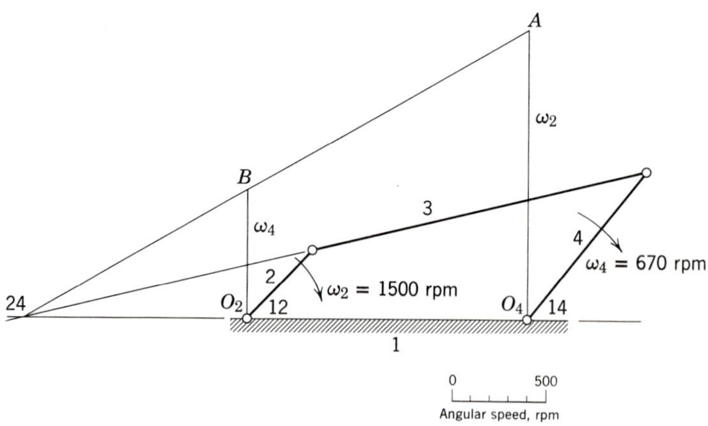

Fig. 3-20. Solution of Example Problem 3-11.

Example 3-11. The crank (link 2) of the four-bar mechanism in Fig. 3-20 is turning at 1500 rpm clockwise. Determine the angular velocity of link 4.

Solution

1. At 14 lay out 14-A, perpendicular to 12-14, and representing ω_2 to the scale shown.
2. Draw a gauge line from A to 24.
3. At 12 erect a line perpendicular to 12-14 and terminating at B on the gauge line. Line 12-B represents ω_4. Applying the scale, $\omega_4 = 670$ rpm.
4. Since 24 lies outside segment 12-14, the links 2 and 4 turn in the same direction. Therefore the angular velocity of link 4 is 670 rpm clockwise.

3-11. Velocity Image Method

Rules. The velocity image method is an application of the relative velocity approach discussed in Chapter 2. In review, the basic rules are:

1. The relative velocity of two moving points is the vector difference of their absolute velocities.
2. The absolute velocity of one point is the sum of the absolute velocity of a second point and the velocity of the first relative to the second.
3. The relative velocity of two points on the same body is perpendicular to the line joining the points.
4. The magnitude of the angular velocity of a body can be found by dividing the magnitude of the relative velocity of any two points on the body by the distance between them.

Notation. In Fig. 3-21*a* a vector triangle has been drawn. In Fig. 3-21*b* the same vector triangle has been reproduced, but using the simplified notation convenient in velocity image analyses. Point *o* is called the *pole*

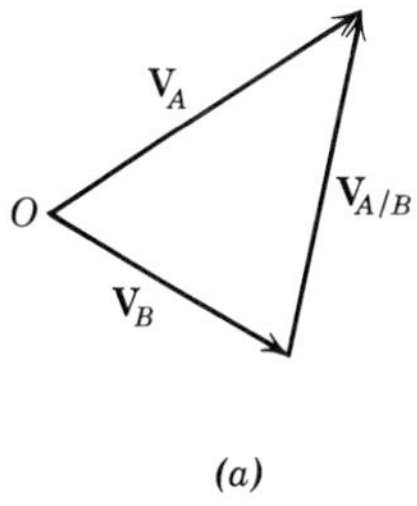

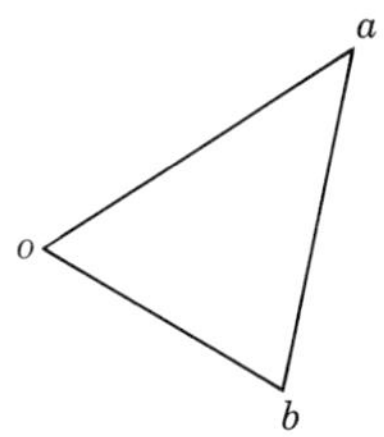

$$\mathbf{V}_A = \vec{oa}$$
$$\mathbf{V}_B = \vec{ob}$$
$$\mathbf{V}_{A/B} = \vec{ba}$$
$$\mathbf{V}_{B/A} = \vec{ab}$$

(*b*)

Fig. 3-21. Notation.

point, and at it are the origins of the absolute velocities. The terminal point of an absolute velocity is marked by a lower-case letter. For example, the terminus of $\mathbf{V}_A$ is a, while that of $\mathbf{V}_B$ is b. The following rules govern the interpretation of the velocity diagram in this simplified form:

1. Absolute velocities originate at the pole point, o, and therefore the sense of each is away from the pole point.
2. Lines joining other lower-case letters represent relative velocities. For example, line ab represents $\mathbf{V}_{A/B}$ or $\mathbf{V}_{B/A}$. The sense of a relative velocity is away from the letter corresponding to the denominator of the subscript in the name of the velocity. For example, the sense of $\mathbf{V}_{B/A}$ is from a toward b, while the sense of $\mathbf{V}_{A/B}$ is from b toward a.

Velocity diagram. The construction of a velocity diagram for a mechanism will be illustrated by an example.

Example 3-12. A four-bar mechanism is illustrated in Fig. 3-22. The velocity of point A is known. The velocities of points B, C, and D, as well as the angular velocities of links 3 and 4, are required.

Solution

1. Locate the pole point, o, for the velocity diagram at any convenient position on the sheet and choose a velocity scale.

2. Lay out the velocity $\mathbf{V}_A$ to scale as oa from the pole point.

3. The vector equation relating the velocities of points A and B is

$$\mathbf{V}_B = \mathbf{V}_A + \mathbf{V}_{B/A}$$

$\mathbf{V}_A$ is completely known. $\mathbf{V}_B$ must be perpendicular to O_4B, since link 4 rotates about O_4. And $\mathbf{V}_{B/A}$ must be perpendicular to line AB since A and B are two points on the same link. Two unknowns only remain, the magnitudes of $\mathbf{V}_B$ and $\mathbf{V}_{B/A}$.

4. Since $\mathbf{V}_B$ is an absolute velocity, it must originate at the pole point. Therefore, draw a line from the pole point, perpendicular to O_4B, and of indefinite length.

5. $\mathbf{V}_{B/A}$ is a relative velocity; thus it will be represented by a line ab in the diagram. Therefore, from a draw a line perpendicular to AB and of indefinite length.

6. The intersection of the two lines just drawn is point b. Connect b to o. Then $\mathbf{V}_B = \vec{ob}$.

7. To find the velocity of point C, vector equations can be written relating the velocities of C and A and of C and B.

$$\mathbf{V}_C = \mathbf{V}_A + \mathbf{V}_{C/A}$$

$$\mathbf{V}_C = \mathbf{V}_B + \mathbf{V}_{C/B}$$

In each equation there are three unknowns: the magnitude and direction of $\mathbf{V}_C$ and the magnitude of the relative velocity. They must be solved simultaneously.

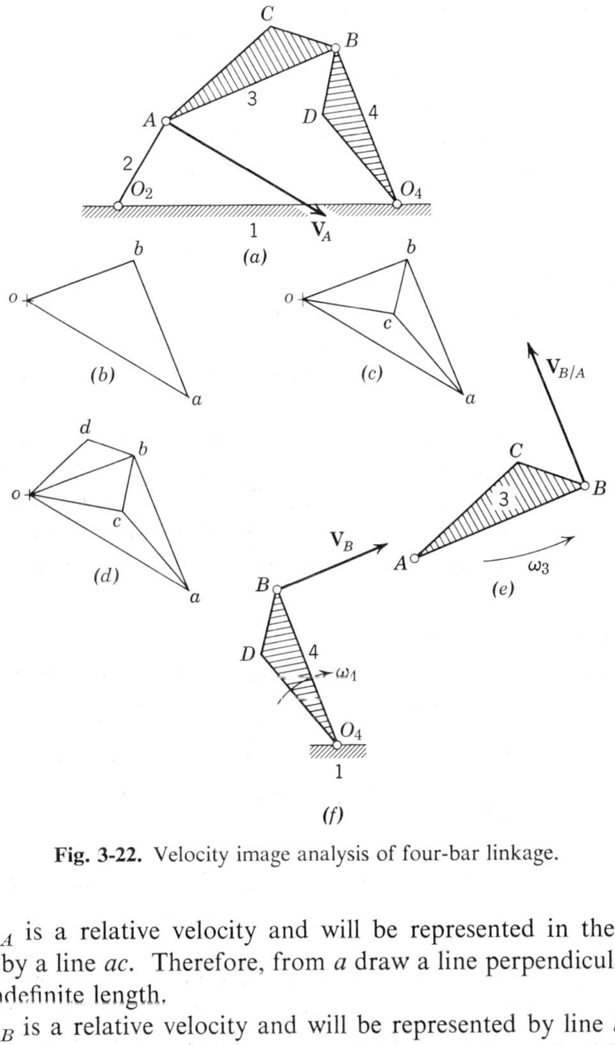

Fig. 3-22. Velocity image analysis of four-bar linkage.

8. $\mathbf{V}_{C/A}$ is a relative velocity and will be represented in the velocity diagram by a line ac. Therefore, from a draw a line perpendicular to CA and of indefinite length.

9. $\mathbf{V}_{C/B}$ is a relative velocity and will be represented by line bc in the diagram. From b draw a line perpendicular to CB and of indefinite length.

10. The intersection of the lines just drawn is point c. Join c to o. Then $\mathbf{V}_C = \vec{oc}$.

11. To find the velocity of D, the following equation can be used:

$$\mathbf{V}_D = \mathbf{V}_B + \mathbf{V}_{D/B}$$

There are only two unknowns, the magnitudes of $\mathbf{V}_D$ and $\mathbf{V}_{D/B}$. $\mathbf{V}_D$ is perpendicular to O_4D and $\mathbf{V}_{D/B}$ to line DB.

12. Since $\mathbf{V}_D$ is an absolute velocity, draw a line from o perpendicular to O_4D and of indefinite length.

13. Since $\mathbf{V}_{D/B}$ is a relative velocity, it will be bd in the diagram. Draw a line from b perpendicular to DB and of indefinite length.

14. The intersection of the lines just drawn is point d. Join d to o. Then $\mathbf{V}_D = \overrightarrow{od}$.

15. The angular velocity of link 3 can now be found. Its magnitude is found by dividing the magnitude of the relative velocity of points B and A by the distance between them. The direction of link 3's rotation can be determined by making a rough sketch (Fig. 3-22*e*) indicating $\mathbf{V}_{B/A}$ at B. This makes it clear that link 3 must be turning in a counterclockwise direction.

16. The angular velocity of link 4 is found in a similar manner. Its magnitude can be found by dividing the magnitude of the velocity of B by its distance from the center of rotation. In Fig. 3-22*f* the link has been sketched and $\mathbf{V}_B$ indicated. The direction of rotation of link 4 can be determined by an inspection of the sketch. Link 4 is turning clockwise.

Image principle. The image principle states:

> The figure formed by the termini of the velocities of three or more points on the same link is similar to the figure formed by the points themselves.

The principle is illustrated in Fig. 3-22. Since ab is perpendicular to AB, bc to BC, and ac to AC, triangles abc and ABC are similar. Points A, B, and C are all on link 3.

The image principle can be useful as a check on the accuracy of the velocity diagram. In Fig. 3-22, for example, after the velocity diagram has been completed (Fig. 3-22*d*), it should be determined whether triangles abc and ABC are, in fact, similar.

A more important use of the principle is to obtain the velocity of other points on a link when the velocities of two points are known. The velocity of C could have been obtained simply by making triangle abc similar to ABC after points a and b had been located.

If C were on line AB, then, when lines were drawn from a and b, perpendicular to AC and BC, it would be found that they coincide, making it impossible to determine an intersection point. However, the location of c

can be found by employing the image principle. Because figures *abc* and *ABC* must be similar, *c* can be located by using the proportion

$$\frac{ca}{ba} = \frac{CA}{BA}$$

The proportion can be solved on the slide rule or graphically.

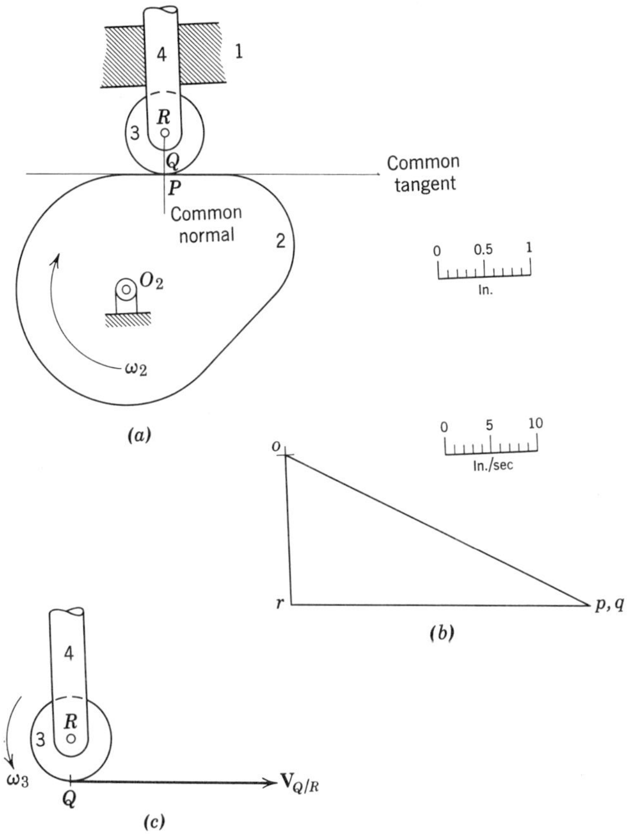

Fig. 3-23. Velocity image analysis of four-link cam mechanism.

Example 3-13. The cam in Fig. 3-23*a* is turning clockwise at 29.45 rad/sec. Determine the velocity of the follower, link 4, and the angular velocity of the roller, link 3.

Solution

1. There is pure rolling between the cam and the roller. Therefore points *P* and *Q* have the same velocity. Select pole point *o* and from it lay out $\mathbf{V}_P$ as *op* (Fig. 3-23*b*). Point *p* is also *q*, since $\mathbf{V}_Q = \mathbf{V}_P$.

2. To find the velocity of the follower, it is sufficient to find the velocity of any point on it. Point R is convenient. The vector equation is

$$\mathbf{V}_R = \mathbf{V}_Q + \mathbf{V}_{R/Q}$$

From o draw a line in the direction of R's motion, and from q a line perpendicular to RQ. Their intersection is r. The follower velocity is 15.4 in./sec at $-90°$.

3. The angular speed of the roller is found by dividing the magnitude of the relative velocity of Q and R by the distance between them.

$$\omega_3 = \frac{32.6 \text{ in./sec}}{0.4 \text{ in.}}$$

$$= 81.5 \text{ rad/sec}$$

The sketch in Fig. 3-23*c* indicates that the roller must be turning counterclockwise. Thus the roller velocity is 81.5 rad/sec counterclockwise.

Example 3-14. A shaper mechanism is shown in Fig. 3-24*a*. Point A_{23} is the center of the pin joint connecting the crank and slider, links 2 and 3. Point A_4 is on link 4 and momentarily coincident with point A_{23}. The crank is turning clockwise, and the speed of point A_{23} is 10 in./sec. Determine the velocity of the ram (link 6) and the angular velocities of links 4 and 5.

Solution

1. First find the velocity of point A_4. The vector equation is

$$\mathbf{V}_{A_4} = \mathbf{V}_{A_{23}} + \mathbf{V}_{A_4/A_{23}}$$

From the pole point in Fig. 3-24*b* the velocity of point A_{23} has been laid out as oa_{23}. From o a line has been drawn in the known direction of the motion of point A_4, and from a_{23} a line has been drawn in the direction of the relative motion of A_{23} and A_4. The relative motion of these points is one of sliding along link 4. These lines intersect at a_4.

2. Point b can now be found by an application of the image principle. Since points O_4, A_4, and B are on the same link, figure oa_4b must be similar to figure O_4A_4B. Therefore,

$$\frac{ob}{oa_4} = \frac{O_4B}{O_4A_4}$$

Solve the proportion and locate b in the velocity diagram. The graphical solution is shown in Fig. 3-24*c*, and point b is located in the velocity diagram in Fig. 3-24*d*.

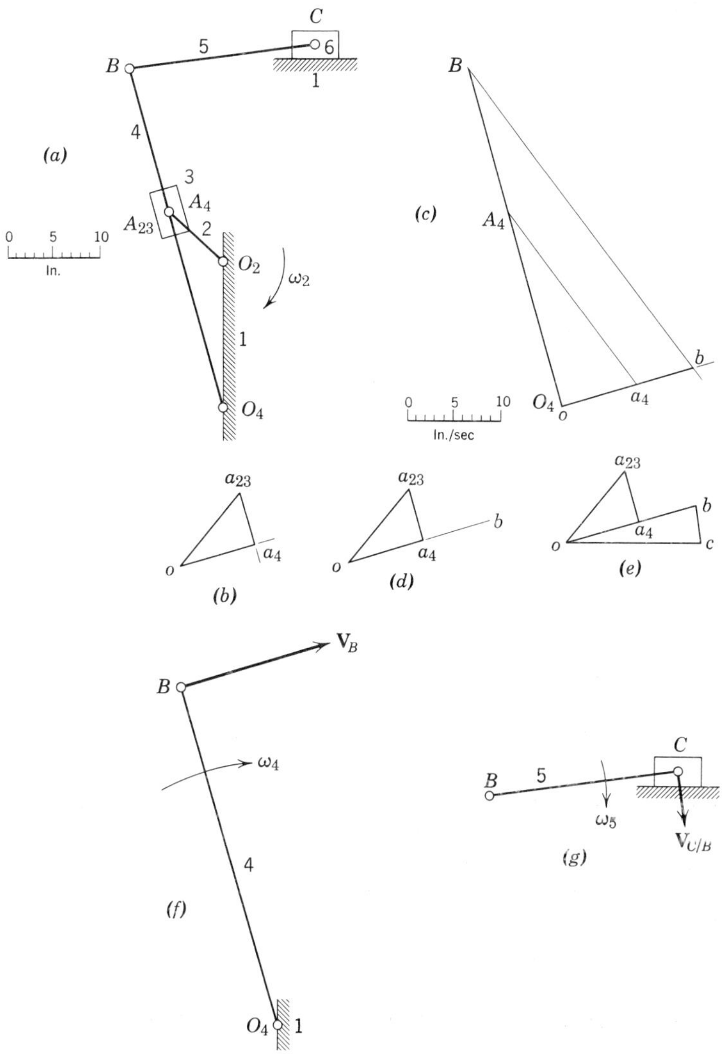

Fig. 3-24. Velocity image analysis of shaper mechanism.

3. Now the velocity of point C can be found, using the vector equation

$$\mathbf{V}_C = \mathbf{V}_B + \mathbf{V}_{C/B}$$

From o a line is drawn in the direction of C's motion, and from b a line is drawn perpendicular to BC. Their intersection point is c. The velocity of the ram is 15 in./sec at 0°.

4. The angular speed of link 4 is

$$\omega_4 = \frac{V_B}{r_4} = \frac{15}{36.5} = 0.41 \text{ rad/sec}$$

The sketch in Fig. 3-24*f* indicates that link 4 is turning clockwise. The angular speed of link 5 is

$$\omega_5 = \frac{V_{C/B}}{r_5} = \frac{4.2}{20.3} = 0.207 \text{ rad/sec}$$

Figure 3-24*g* shows its direction.

3-12. Indirect Method

Figure 3-25*a* illustrates a Watt's walking beam mechanism. Since it is the velocity of point A which is given, the velocity image method will not permit a direct solution. The reason is that, when vector equations are written relating the velocities of other points to that of A, it is found that they have too many unknowns and cannot be solved. Yet, if the velocity of point C were known instead, a solution could be made.

The indirect method is one of several techniques for resolving this dilemma. It consists in assuming the velocity of a point from which the velocity analysis can be made, constructing the velocity diagram, and, as a final step, calculating the scale to which this arbitrarily sized velocity diagram represents the actual velocities.

Inspection of the mechanism in Fig. 3-25*a* leads to the guess that link 5 is turning counterclockwise. From the pole point a line oc of arbitrary length is drawn to represent the velocity of point C. It is, of course, drawn perpendicular to O_5C and with the assumed sense. Then the velocity diagram is constructed by the usual methods. In Fig. 3-25*b* point d has been located, as well as point b in Fig. 3-25*c*, and, finally, point a in Fig. 3-25*d*.

This velocity diagram represents the velocities of points in the mechanism

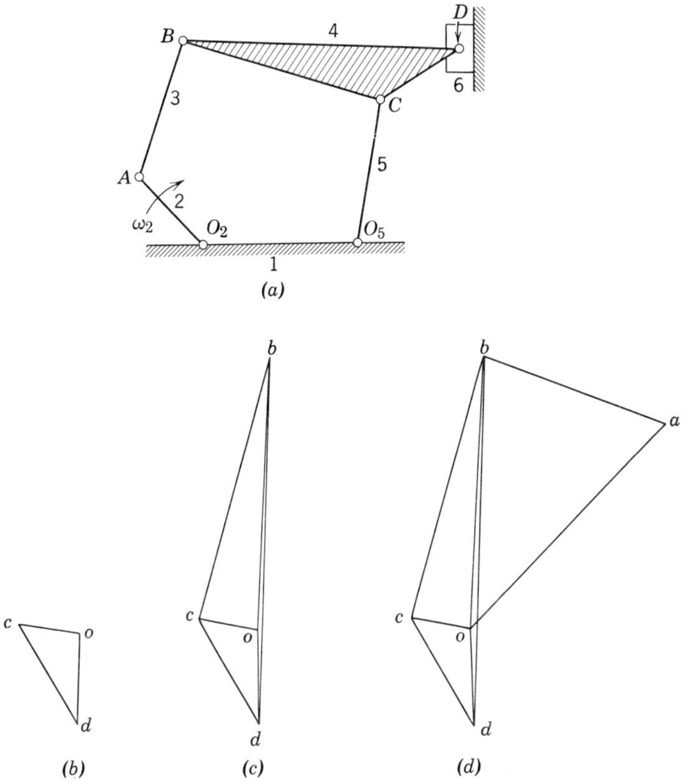

Fig. 3-25. Velocity analysis of Watt's walking beam mechanism: indirect method.

to some undetermined scale. To find the scale, note that V_A, which is known, is represented by drawing distance *oa*, which can be measured. Then

$$k_v = \frac{V_A}{oa}$$

Now that the velocity scale is known, the diagram can be interpreted.

If the assumed sense of $\mathbf{V}_C$ was incorrect, the sense of $\mathbf{V}_A$ in the diagram will be wrong. Then the diagram would have to be redrawn. Often, however, an inspection of the mechanism layout permits an accurate guess.

PROBLEMS

3-1. Locate all of the instant centers for the mechanism shown in Fig. P3-1.
3-2. Locate all of the instant centers for the mechanism shown in Fig. P3-2.

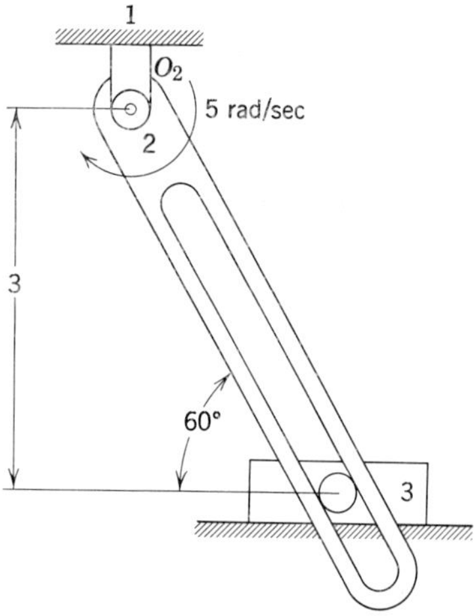

Fig. P3-1

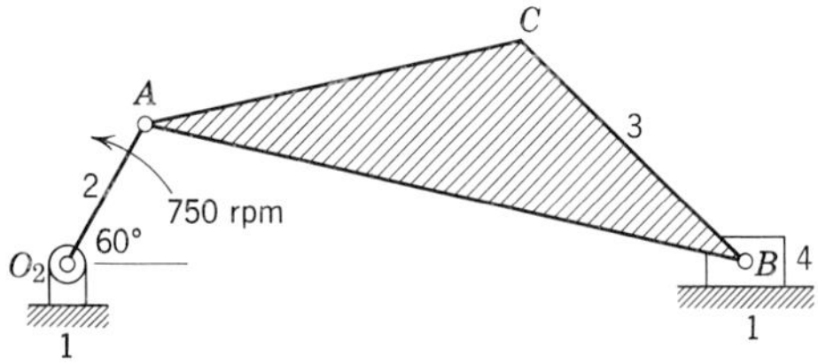

Fig. P3-2

3-3. Locate all of the instant centers for the mechanism shown in Fig. P3-3.
3-4. Locate all of the instant centers for the mechanism shown in Fig. P3-4.

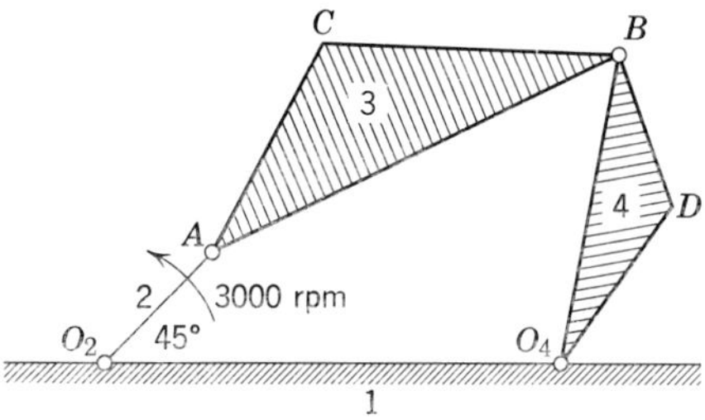

$O_2O_4 = 3''$; $O_2A = 1''$; $AB = 3''$; $O_4B = 2''$;
$AC = 1\frac{1}{2}''$; $BC = 2''$; $O_4D = 1\frac{1}{4}''$; $BD = 1''$

Fig. P3-3

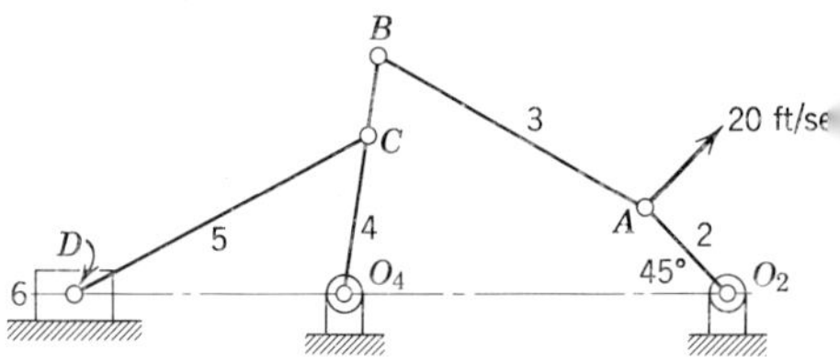

$O_2O_4 = 2.5''$; $O_2A = 0.75''$; $AB = 2''$
$O_4B = 1.5''$; $O_4C = 1''$; $CD = 2.5''$

Fig. P3-4

3-5. Locate all of the instant centers for the mechanism shown in Fig. P3-5.

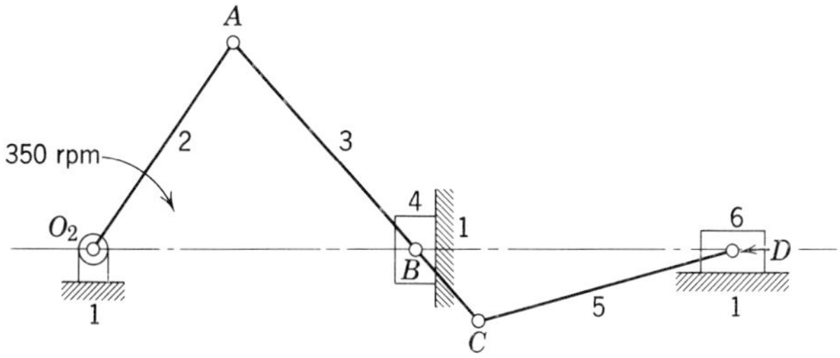

$O_2B = 2.5''$; $BD = 2.5''$; $O_2A = 1\frac{7}{8}''$; $AB = 2\frac{1}{16}''$; $CD = 2\frac{1}{16}''$

Fig. P3-5

3-6. Locate all of the instant centers for the mechanism shown in Fig. P3-6.

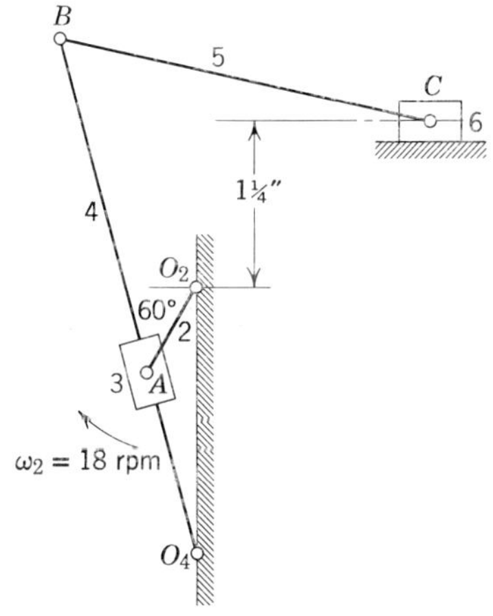

$O_2O_4 = 2''$; $O_2A = \frac{3}{4}''$; $O_4B = 4''$; $BC = 3''$

Fig. P3-6

3-7. Locate all of the instant centers for the mechanism shown in Fig. P3-7.

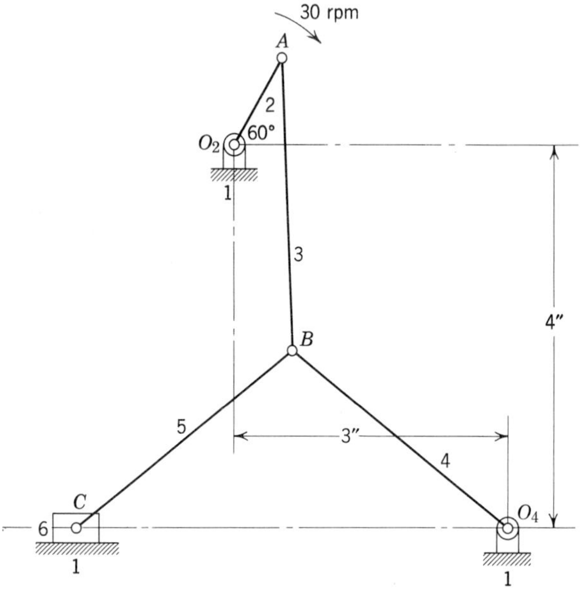

$O_2A = 1''$; $AB = BC = BO_4 = 3''$

Fig. P3-7

3-8. Locate all of the instant centers for the mechanism shown in Fig. P3-8.

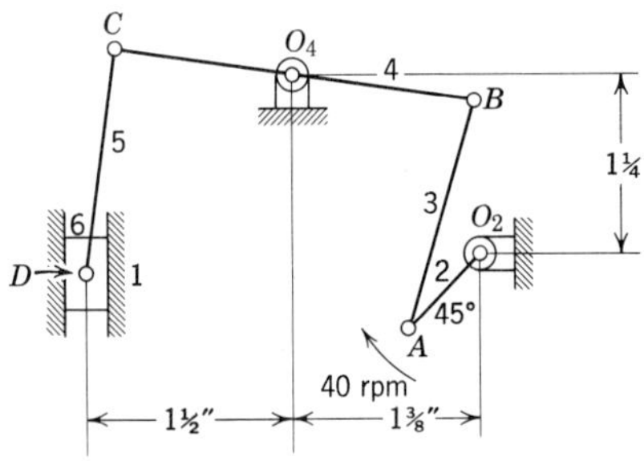

$O_2A = 23/32''$; $AB = 1\tfrac{5}{8}''$; $BO_4 = O_4C = 1\tfrac{5}{16}''$; $CD = 1\tfrac{7}{16}''$

Fig. P3-8

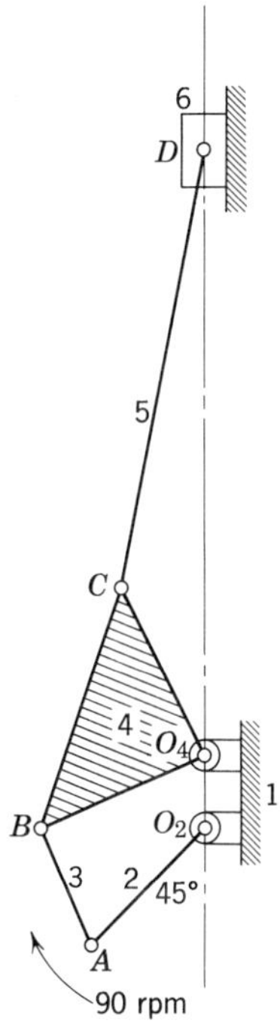

$O_2O_4 = \frac{1}{2}''$; $O_2A = 1\frac{1}{8}''$; $AB = \frac{7}{8}''$; $BO_4 = 1\frac{1}{4}''$; $O_4C = 1\frac{1}{4}''$; $BC = 1\frac{3}{4}''$; $CD = 3''$

Fig. P3-9

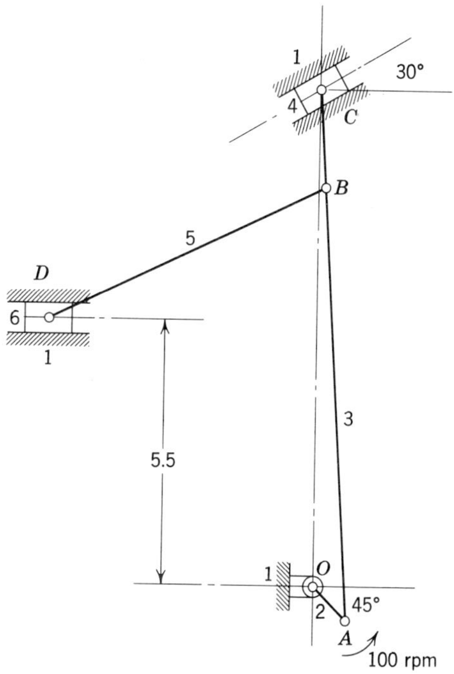

$OA = 2''$; $AC = 11''$; $AB = 9''$; $BD = 7.5''$

Fig. P3-10

3-9. Locate all of the instant centers for the mechanism shown in Fig. P3-9.

3-10. Locate all of the instant centers for the mechanism shown in Fig. P3-10.

3-11. Using instant center methods, determine the velocity of the slider in the mechanism of Fig. P3-1.

3-12. Using instant center methods, determine the velocities of points *B* and *C*, and of link 3 in the slider-crank mechanism of Fig. P3-2.

3-13. Using instant center methods, determine the velocities of points *B*, *C*, and *D*, and of links 3 and 4 in the four-bar mechanism of Fig. P3-3.

3-14. Using instant center methods, determine the velocities of points *B*, *C*, and *D*, and of links 3, 4, and 5 in the crank and rocker quick-return mechanism of Fig. P3-4.

3-15. Using instant center methods, determine the velocities of points *B*, *C*, and *D*, and of links 3 and 5 in the double-slider mechanism of Fig. P3-5.

3-16. Using instant center methods, determine the velocities of points *B* and *C*, and of links 4 and 5 in the quick-return mechanism of Fig. P3-6.

3-17. Using instant center methods, determine the velocities of points *B* and *C*, and of links 3, 4, and 5 in the toggle mechanism of Fig. P3-7.

3-18. Using instant center methods, determine the velocities of points *B*, *C*, and *D*, and of links 3, 4, and 5 in the pumping mechanism of Fig. P3-8.

3-19. Using instant center methods, determine the velocities of points *B*, *C*, and *D*, and of links 3, 4, and 5 in the drag-link, quick-return mechanism of Fig. P3-9.

3-20. Using instant center methods, determine the velocities of points *B*, *C*, and *D*, and of links 3 and 5 in the Marshall valve gear mechanism of Fig. P3-10.

3-21. Solve problem 3-11 by velocity image methods.

3-22. Solve problem 3-12 by velocity image methods.

3-23. Solve problem 3-13 by velocity image methods.

3-24. Solve problem 3-14 by velocity image methods.

3-25. Solve problem 3-15 by velocity image methods.

3-26. Solve problem 3-16 by velocity image methods.

3-27. Solve problem 3-17 by velocity image methods.

3-28. Solve problem 3-18 by velocity image methods.

3-29. Solve problem 3-19 by velocity image methods.

3-30. Solve problem 3-20 by velocity image methods.

4 Acceleration Analysis

4-1. Introduction

Accelerations interest designers chiefly because of the stresses that accompany them. If the links of a mechanism are to be given the linear and angular accelerations they are to experience in their normal cycle of motion, forces and moments have to be applied. These forces and moments cause stresses on the members in addition to the stresses caused by the working loads. In modern machinery, with its ever-increasing speeds of operation, higher and higher accelerations are being called for. And the stresses that accompany these higher accelerations are substantially greater, in many instances, than those caused by the working loads. A stress analysis that overlooks these dynamic stresses will dangerously underestimate the strength requirements of the links.

As the general method for acceleration analyses of mechanisms, an acceleration image technique will be presented in this chapter which is similar in approach to the velocity image technique. Again, as in velocity analysis, the analysis is carried out for the mechanism at a particular phase. It is assumed that for any phase the mechanism's geometry and the acceleration of some point or link are known. A complete acceleration analysis requires a repetition of the procedure for each phase in which the designer is interested.

4-2. Relative Acceleration of Two Points on a Link

Basic rules. The acceleration of a point moving on a circular path about a fixed center has been dealt with in Art. 2-6. In order to carry out the acceleration analysis of a mechanism, it is necessary to argue from the known acceleration of some point in the mechanism to the acceleration of some second point. In this article the relationship between theaccelerations of two points on the same link will be discussed. In a later article the relationship between the accelerations of two points on different links will be examined.

Points A and B in Fig. 4-1a are points on a rigid link which has plane motion. The rule relating the accelerations of points A and B is similar to that relating their velocities.

> The acceleration of a point in a mechanism is equal to the vector sum of the acceleration of any second point and the acceleration of the first relative to the second.

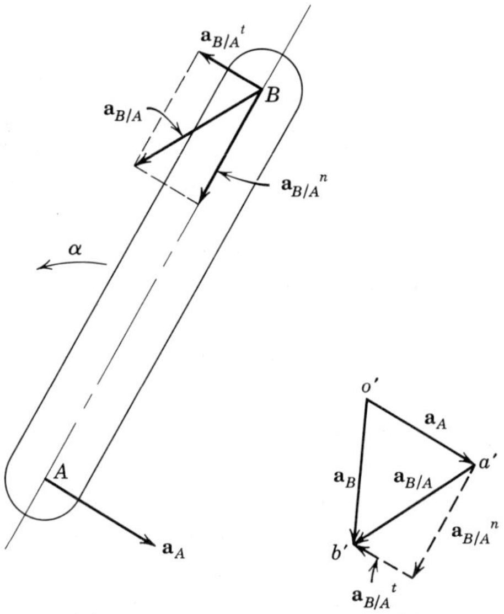

Fig. 4-1. Relative acceleration of two points on a rigid link.

This rule in equation form, where A and B are any two points, is

$$\mathbf{a}_B = \mathbf{a}_A + \mathbf{a}_{B/A} \tag{4-1}$$

It is most convenient to resolve the relative acceleration into components which can be determined separately. For any two points, A and B, *on the same link*,

$$\mathbf{a}_{B/A} = \mathbf{a}^n_{B/A} + \mathbf{a}^t_{B/A} \tag{4-2}$$

The sense of the normal component, $\mathbf{a}^n_{B/A}$, is from B and toward A. The tangential component. $\mathbf{a}^t_{B/A}$, is perpendicular to line BA and with a sense consistent with the sense of the angular acceleration of the link.

The equations for the magnitudes of the normal and tangential components of the relative acceleration are similar to those for the acceleration of a point moving on a circular path about a fixed center. The acceleration of B relative to A is the acceleration B has moving in a circular path about A as center. Thus,

$$a^n_{B/A} = \frac{V^2_{B/A}}{r_{BA}} \tag{4-3}$$

and

$$a^t_{B/A} = r_{BA} \cdot \alpha \tag{4-4}$$

Equation 4-3 is similar to Eq. 2-14a, and Eq. 4-4 to Eq. 2-15.

Equation 4-4 can be rewritten in a form that will be useful in calculating the angular accelerations of links.

$$\alpha = \frac{a^t_{B/A}}{r_{BA}} \tag{4-4a}$$

Equation 4-4a can be stated as a rule:

> To find the magnitude of the angular acceleration of a link, divide the magnitude of the tangential component of the relative acceleration of any two points on the link by the distance between them.

In Eqs. 4-3 to 4-4a, r_{BA} is the actual distance on the link between points A and B. AB (or BA) is the drawing distance.

Image principle. The acceleration analysis of a mechanism proceeds by constructing an acceleration diagram in a manner analogous to the construction of the velocity diagram. And just as in the velocity diagram, an image principle applies.

> In the acceleration diagram, the figure formed by the termini of the acceleration vectors for three or more points on the same link is similar to the figure formed by the points themselves.

4-3. Examples of Acceleration Analysis

The acceleration diagram for a mechanism is more complex than the velocity diagram. Therefore a more detailed notation will be used. The rules for notation, illustrated in Fig. 4-1b, are:

1. The pole point is o'.
2. Termini of absolute accelerations are designated by primed lower-case letters. For example, the terminus of $\mathbf{a}_B$ is b'.
3. The normal and tangential components of accelerations are shown as broken lines.
4. Arrowheads will be used to indicate the sense of vectors unambiguously, and the vectors will usually be labeled.

The procedures used in constructing acceleration diagrams will be illustrated by two examples.

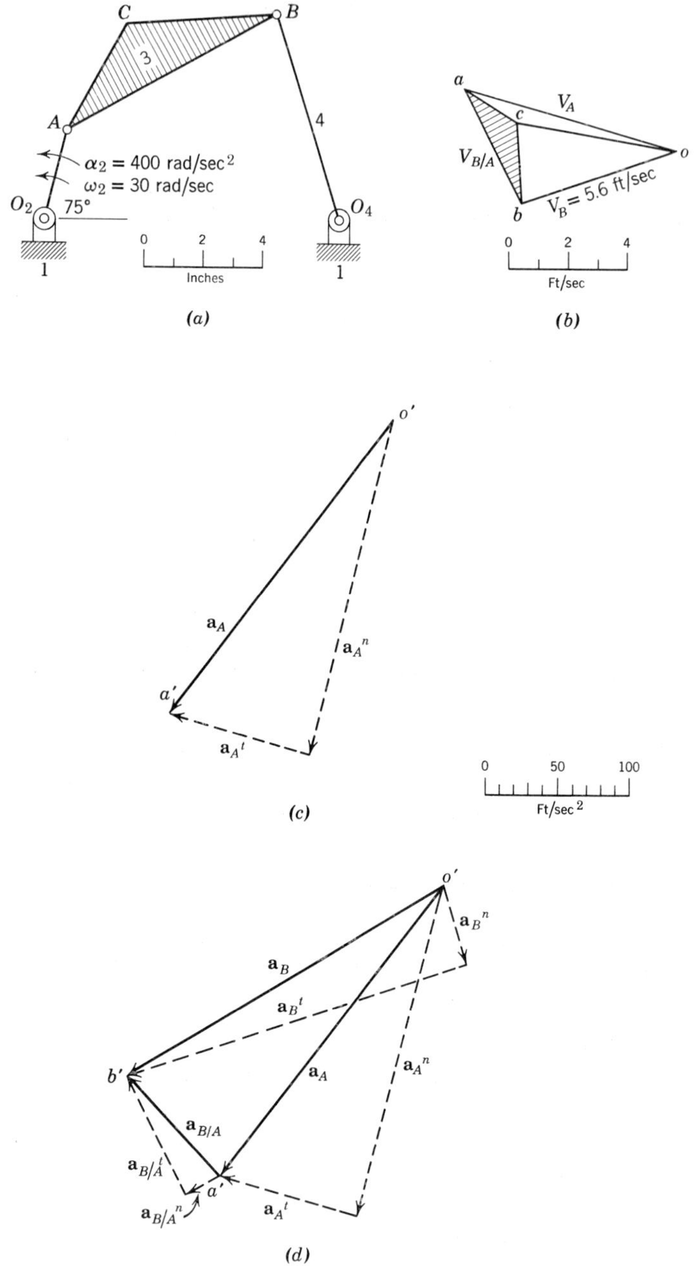
C
B
3
A
4
$\alpha_2 = 400$ rad/sec^2
$\omega_2 = 30$ rad/sec
O_2
75°
O_4
1
1
0 2 4
Inches
(a)
a
V_A
c
$V_{B/A}$
o
b
$V_B = 5.6$ ft/sec
0 2 4
Ft/sec
(b)
o′
$\mathbf{a}_A$
$\mathbf{a}_A{}^n$
a′
$\mathbf{a}_A{}^t$
0 50 100
Ft/sec^2
(c)
o′
$\mathbf{a}_B{}^n$
$\mathbf{a}_B$
$\mathbf{a}_B{}^t$
b′
$\mathbf{a}_A$
$\mathbf{a}_A{}^n$
$\mathbf{a}_{B/A}$
$\mathbf{a}_{B/A}{}^t$
a′
$\mathbf{a}_{B/A}{}^n$
$\mathbf{a}_A{}^t$
(d)

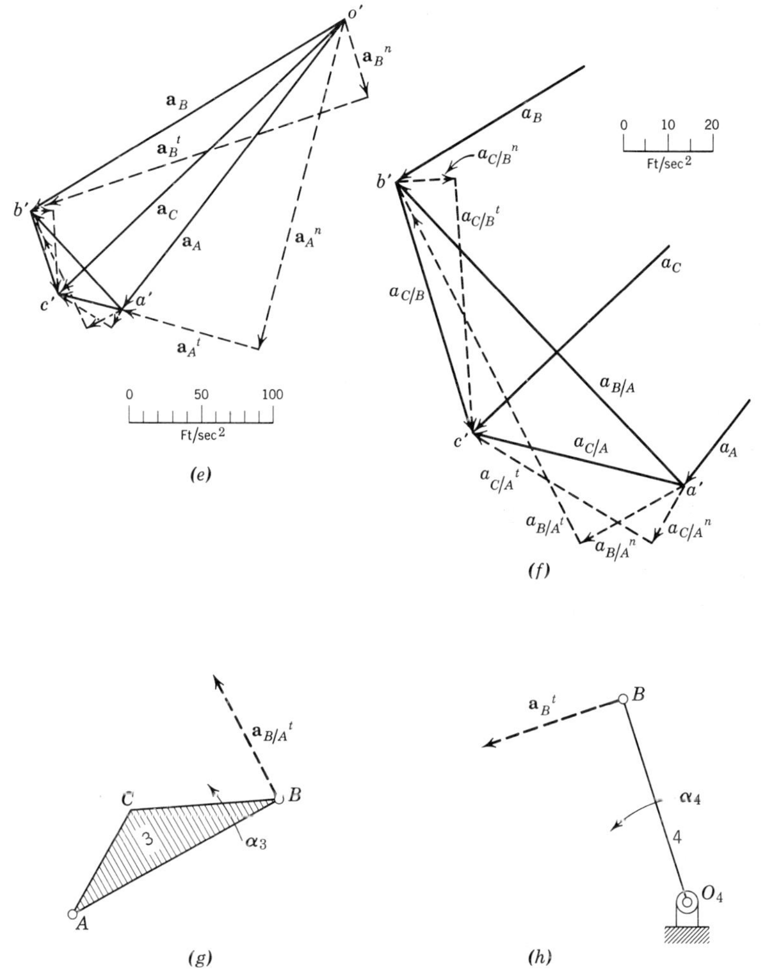

Fig. 4-2. Acceleration analysis of a four-bar mechanism.

Example 4-1

A four-bar mechanism is shown in Fig. 4-2*a*. Find the accelerations of points B and C, as well as the angular accelerations of links 3 and 4.

Solution. The velocity diagram is needed in order to furnish information that will be used in calculating normal components of accelerations. The completed velocity diagram is shown in Fig. 4-2*b*.

To find $\mathbf{a}_A$:

1. Applying Eqs. 2-14 and 2-15,

$$\begin{aligned} a_A{}^n &= r\omega^2 \\ &= (3/12)(30)^2 \\ &= 225 \text{ ft/sec}^2 \\ a_A{}^t &= r_A \cdot \alpha_2 \\ &= (3/12)(400) \\ &= 100 \text{ ft/sec}^2 \end{aligned}$$

2. The normal component, $\mathbf{a}_A{}^n$, is from A toward O_2. It has been drawn to scale from o' in Fig. 4-2*c*.

3. The tangential component, $\mathbf{a}_A{}^t$, is laid out from the terminus of, and perpendicular to, the normal component. Since the angular acceleration of link 2 is counterclockwise, the tangential component is up and to the left.

4. The acceleration of point A, $\mathbf{a}_A$, is the vector from o' to the terminus of $\mathbf{a}_A{}^t$. Its terminus is a'.

To find $\mathbf{a}_B$:

1. The vector equation relating the accelerations of points A and B is

$$\mathbf{a}_B = \mathbf{a}_A + \mathbf{a}_{B/A}$$

And since points A and B are on the same link,

$$\mathbf{a}_B = \mathbf{a}_A + \mathbf{a}_{B/A}^n + \mathbf{a}_{B/A}^t$$

There are three unknowns in this equation: the magnitude of $\mathbf{a}_{B/A}^t$ and the magnitude and direction of $\mathbf{a}_B$. The equation cannot, therefore, be solved as it stands. However, since point B revolves about a fixed center, O_4, Eq. 2-16 applies.

$$\mathbf{a}_B = \mathbf{a}_B{}^n + \mathbf{a}_B{}^t$$

The acceleration of B can be found by the simultaneous solution of these two vector equations.

2. The magnitudes of the normal components are

$$\begin{aligned} a_B{}^n &= \frac{V_B{}^2}{r_B} & \qquad a_{B/A}^n &= \frac{V_{B/A}^2}{r_{BA}} \\ &= \frac{(5.6)^2}{(7/12)} & &= \frac{(4.18)^2}{(8/12)} \\ &= 53.8 \text{ ft/sec}^2 & &= 26.2 \text{ ft/sec}^2 \end{aligned}$$

3. In Fig. 4-2*d*, $\mathbf{a}_B{}^n$ has been laid out from o'. Its sense is from B toward O_4.

4. In the same figure $\mathbf{a}_B{}^t$ has been laid out from the terminus of and perpendicular to $\mathbf{a}_B{}^n$. It is drawn of indefinite length, since its magnitude is unknown.

5. From a' in Fig. 4-2*d* $\mathbf{a}^n_{B/A}$ has been laid out. Its sense is from B toward A.

6. From the terminus of $\mathbf{a}^n_{B/A}$ a line of indefinite length has been drawn perpendicular to the normal component to represent $\mathbf{a}^t_{B/A}$.

7. The intersection of the two tangential components, $\mathbf{a}_B{}^t$ and $\mathbf{a}^t_{B/A}$, is point b', the terminus of $\mathbf{a}_B$. From o' a vector has been drawn to b', representing $\mathbf{a}_B$. Its sense is away from o'.

To find $\mathbf{a}_C$:

1. A vector equation can be written relating the accelerations of points C and A.

$$\mathbf{a}_C = \mathbf{a}_A + \mathbf{a}_{C/A}$$

And, since A and C are on the same link,

$$\mathbf{a}_C = \mathbf{a}_A + \mathbf{a}^n_{C/A} + \mathbf{a}^t_{C/A}$$

But this equation has three unknowns and cannot be solved by itself. Since the acceleration of point B on the same link is known, it is worthwhile to write an equation relating the accelerations of points C and B.

$$\mathbf{a}_C = \mathbf{a}_B + \mathbf{a}^n_{C/B} + \mathbf{a}^t_{C/B}$$

A simultaneous solution of these last two vector equations will provide $\mathbf{a}_C$.

2. The magnitudes of the normal components are

$$a^n_{C/A} = \frac{V^2_{C/A}}{r_{CA}} = \frac{(2.15)^2}{(4/12)} = 13.9 \text{ ft/sec}^2$$

$$a^n_{C/B} = \frac{V^2_{C/B}}{r_{CB}} = \frac{(2.6)^2}{(5/12)} = 16.2 \text{ ft/sec}^2$$

3. From the terminus of $\mathbf{a}_A$, the normal component $\mathbf{a}^n_{C/A}$ has been drawn. Its sense is from C toward A.

4. From the terminus of $\mathbf{a}^n_{C/A}$, a line has been drawn perpendicular to it and of indefinite length to represent $\mathbf{a}^t_{C/A}$.

5. From the terminus of $\mathbf{a}_B$, the normal component $\mathbf{a}^n_{C/B}$ has been drawn. Its sense is from C toward B.

6. From the terminus of and perpendicular to $\mathbf{a}^n_{C/B}$, a line has been drawn of indefinite length to represent $\mathbf{a}^t_{C/B}$.

7. The intersection of $\mathbf{a}^t_{C/A}$ and $\mathbf{a}^t_{C/B}$ is point c', the terminus of $\mathbf{a}_C$. Therefore $\mathbf{a}_C$ is represented in the acceleration diagram by the vector from o' to c'.

The image principle can now be applied as a check on the accuracy of the construction in the acceleration diagram. Since points A, B, and C are on the same link, triangles ABC and $a'b'c'$ must be similar. A portion of the diagram has been enlarged in Fig. 4-2*f*.

To find the angular accelerations:

1. Applying Eqs. 2-15 and 4-4*a* to links 3 and 4,

$$\alpha_3 = \frac{a^t_{B/A}}{r_{BA}} = \frac{88}{(8/12)} = 132 \text{ rad/sec}^2$$

$$\alpha_4 = \frac{a_B{}^t}{r_B} = \frac{246}{(7/12)} = 422 \text{ rad/sec}^2$$

2. To determine the direction of the angular acceleration of a link, a rough sketch of the link can be made with the tangential component of the acceleration of a point on the link, or of the relative acceleration of two points on the link, indicated. The senses of the tangential component and the angular acceleration must be consistent. In Fig. 4-2*g*, link 3 has been sketched (not to scale), and the tangential component $\mathbf{a}^t_{B/A}$ has been shown (also not to scale). The sense of this component, regarding B as revolving about A, implies that link 3 is accelerating in a counterclockwise direction.

3. In Fig. 4-2*h* link 4 has been sketched. Point B on this link revolves about fixed center O_4. The tangential component of B's acceleration, $\mathbf{a}_B{}^t$, has been drawn. Its sense implies that the link has a counterclockwise acceleration.

Example 4-2

For the toggle mechanism in Fig. 4-3*a*, find the accelerations of points *A*, *B*, *C*, and *D*, and the angular accelerations of links 3, 4, and 5.

Solution. The velocity diagram has been constructed in Fig. 4-3*b*. Point *d* was found by the image principle, using the proportion

$$\frac{ad}{ab} = \frac{AD}{AB}$$

To find $\mathbf{a}_A$:

1. Since *A* revolves about a fixed center,

$$\mathbf{a}_A = \mathbf{a}_A{}^n + \mathbf{a}_A{}^t$$

2. The magnitude of the normal component is

$$\begin{aligned} a_A{}^n &= \frac{V_A{}^2}{r_A} \\ &= \frac{(12.57)^2}{(2/12)} \\ &= 945 \text{ ft/sec}^2 \end{aligned}$$

3. Since $\alpha_2 = 0$, $a_A{}^t = 0$, and $\mathbf{a}_A = \mathbf{a}_A{}^n$. The acceleration of *A* has been laid off to scale from o' in Fig. 4-3*c*.

To find $\mathbf{a}_B$:

1. Since points *B* and *A* are on the same link,

$$\mathbf{a}_B = \mathbf{a}_A + \mathbf{a}_{B/A}^n + \mathbf{a}_{B/A}^t$$

And since *B* revolves about a fixed center,

$$\mathbf{a}_B = \mathbf{a}_B{}^n + \mathbf{a}_B{}^t$$

These vector equations can be solved simultaneously for the acceleration of point *B*.

2. The magnitudes of the normal components are

$$\begin{aligned} a_B{}^n &= \frac{V_B{}^2}{r_B} \\ &= \frac{(11.4)^2}{(6/12)} \\ &= 260 \text{ ft/sec}^2 \\ a_{B/A}^n &= \frac{V_{B/A}^2}{r_{BA}} \\ &= \frac{(8.5)^2}{(6/12)} \\ &= 144 \text{ ft/sec}^2 \end{aligned}$$

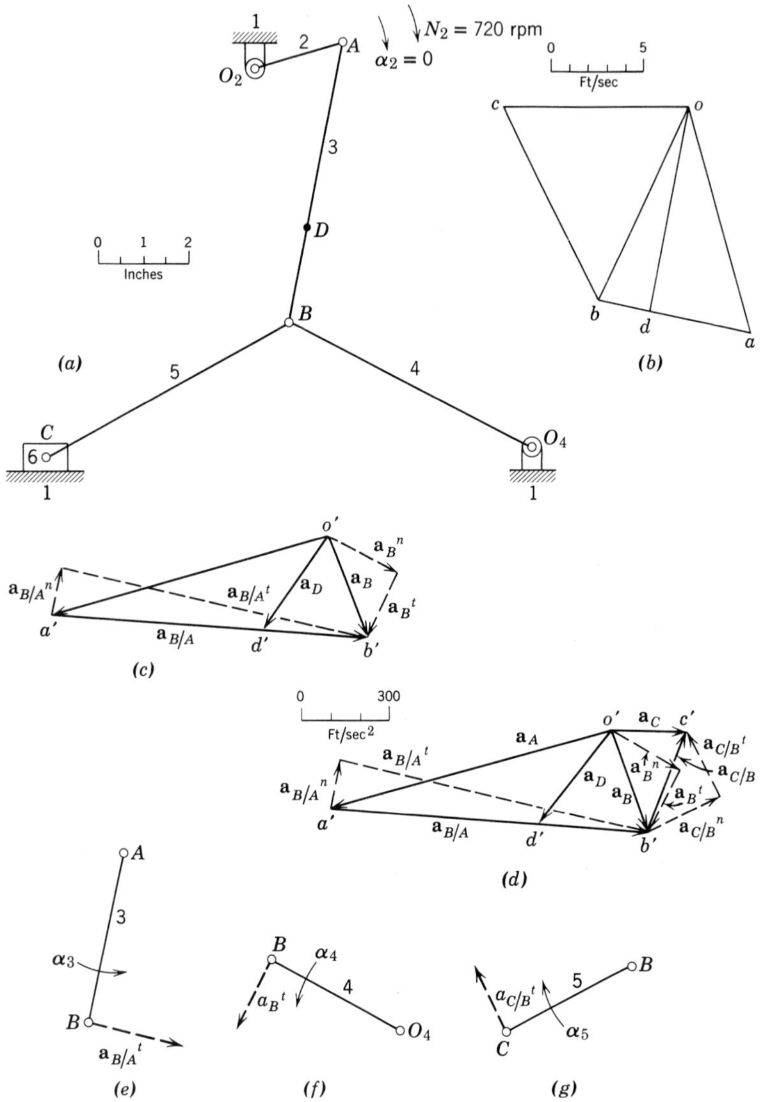

Fig. 4-3. Acceleration analysis of a toggle mechanism.

3. In Fig. 4-3*c* $\mathbf{a}_B{}^n$ has been laid out from o'. Its sense is toward O_4.

4. From the terminus of $\mathbf{a}_B{}^n$, and perpendicular to it, a line of indefinite length has been drawn to represent $\mathbf{a}_B{}^t$.

5. From a', the terminus of $\mathbf{a}_A$, the normal component, $\mathbf{a}^n_{B/A}$, of the relative acceleration of B to A has been drawn. Its sense is from B toward A.

6. And from the terminus of and perpendicular to this normal component has been drawn a line of indefinite length to represent the tangential component, $\mathbf{a}^t_{B/A}$.

7. The intersection of the tangential components is b', the terminus of $\mathbf{a}_B$. The acceleration of B is represented in the acceleration diagram, therefore, by a vector drawn from o' to b'.

To find $\mathbf{a}_D$:

1. Since point D is on the same link as points A and B, whose accelerations are already known, the normal procedure would be to solve simultaneously two vector equations, one relating the accelerations of points D and A, and the other those of points D and B. The equations are

$$\mathbf{a}_D = \mathbf{a}_A + \mathbf{a}^n_{D/A} + \mathbf{a}^t_{D/A}$$

$$\mathbf{a}_D = \mathbf{a}_B + \mathbf{a}^n_{D/B} + \mathbf{a}^t_{D/B}$$

However, since points A, B, and D are collinear, the tangential components, $\mathbf{a}^t_{D/A}$ and $\mathbf{a}^t_{D/B}$, will each be perpendicular to line AB. They will be collinear in the acceleration diagram, making it impossible to determine their intersection point.

2. Therefore the image principle must be used. Since points A, B, and D are on the same link, figures ADB and $a'd'b'$ must be similar. Point d' can be located in the acceleration diagram along line $a'b'$ by determining distance $a'd'$ from the following proportion:

$$\frac{a'd'}{a'b'} = \frac{AD}{AB}$$

This proportion can be solved graphically or by slide rule.

To find $\mathbf{a}_C$:

1. Points B and C are on the same link. Therefore,

$$\mathbf{a}_C = \mathbf{a}_B + \mathbf{a}^n_{C/B} + \mathbf{a}^t_{C/B}$$

This vector equation can be solved immediately, for there are only two unknowns: the magnitudes of $\mathbf{a}_C$ and $\mathbf{a}^t_{C/B}$. The direction of $\mathbf{a}_C$ is necessarily horizontal: point C is on a slider moving horizontally.

2. The magnitude of $\mathbf{a}^n_{C/B}$ is:

$$a^n_{C/B} = \frac{V^2_{C/B}}{r_{CB}}$$

$$= \frac{(11.4)^2}{(6/12)}$$

$$= 260 \text{ ft/sec}^2$$

3. In Fig. 4-3d the normal component, $\mathbf{a}^n_{C/B}$, has been drawn from b', the terminus of $\mathbf{a}_B$.

4. And from the terminus of and perpendicular to $\mathbf{a}^n_{C/B}$ a line has been drawn of indefinite length to represent $\mathbf{a}^t_{C/B}$.

5. From o' a line of indefinite length has been drawn in the direction of the acceleration of C, horizontally.

6. The intersection of $\mathbf{a}_C$ and $\mathbf{a}^t_{C/B}$ is c', the terminus of $\mathbf{a}_C$. The vectors $\mathbf{a}_C$ and $\mathbf{a}_{C/B}$ can now be drawn in the acceleration diagram as $\overrightarrow{o'c'}$ and $\overrightarrow{b'c'}$.

To find the angular accelerations:

1. The magnitudes of the angular accelerations of links 3, 4, and 5 are

$$\alpha_3 = \frac{a^t_{B/A}}{r_{BA}}$$

$$= \frac{(1025)}{(6/12)}$$

$$= 2050 \text{ rad/sec}^2$$

$$\alpha_4 = \frac{a_B{}^t}{r_B}$$

$$= \frac{(230)}{(6/12)}$$

$$= 460 \text{ rad/sec}^2$$

$$\alpha_5 = \frac{a^t_{C/B}}{r_{CB}}$$

$$= \frac{(230)}{(6/12)}$$

$$= 460 \text{ rad/sec}^2$$

2. Sketches (not to scale) of the links in Figs. 4-3e, f, g, with tangential components of accelerations indicated, suffice to determine the directions

of the angular accelerations. Once the sketches are made, the directions of the angular accelerations are determined by inspection. They have been indicated in the sketches.

4-4. Graphical Determination of Normal Acceleration

Point A in Fig. 4-4*a* is revolving about a fixed center, O. It is required to find the normal component of its acceleration, $\mathbf{a}_A{}^n$. The following procedure can be used.

1. At A lay out to scale the velocity of A, $\mathbf{V}_A$.
2. Call the terminus of $\mathbf{V}_A$ 'V' and connect it to O.
3. At V draw a line perpendicular to OV and intersecting OA produced at N.

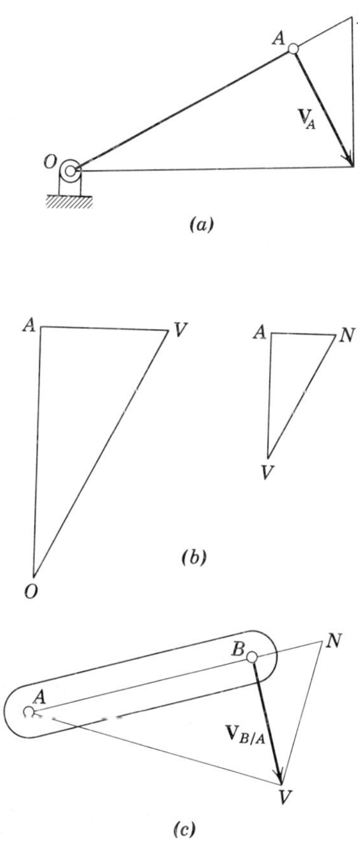

Fig. 4-4. Graphical determination of normal acceleration.

4. Then line AN represents the magnitude of $\mathbf{a}_B{}^n$ to the scale k_a, where

$$k_a = \frac{k_v^2}{k_s} \tag{4-5}$$

k_s = distance scale, ft/in. or in./in.

k_v = velocity scale, ft/sec/in. or in./sec/in.

k_a = acceleration scale, ft/sec²/in. or in./sec²/in.

Proof

1. Since triangles OAV and VAN are similar (Fig. 4-4*b*),

$$\frac{OA}{AV} = \frac{AV}{AN}$$

or

$$AN = \frac{(AV)^2}{OA}$$

2. The magnitude of the normal acceleration of point A is

$$a_A{}^n = \frac{V_A^2}{r_A}$$

3. Since OA represents r_A and AV represents V_A,

$$r_A = k_s \cdot OA$$
$$V_A = k_v \cdot AV$$

4. Substituting,

$$a_A{}^n = \frac{k_v^2(AV)^2}{k_s \cdot OA}$$

5. Since $AN = (AV)^2/OA$,

$$a_A{}^n = \frac{k_v^2}{k_s} AN$$

6. Therefore AN represents $a_A{}^n$ to the scale k_a, where $k_a = k_v^2/k_s$.

Any two of the three scales in Eq. 4-5 can be selected arbitrarily and the third calculated. Since the mechanism layout and the velocity diagram must be drawn before the acceleration analysis can be made, it is usual to select the distance and velocity scales arbitrarily and to calculate the acceleration scale.

Because the direction of the normal acceleration is never in doubt, no rule will be given regarding it in the construction. Generally the construction is used merely to determine the magnitude of the normal component of acceleration.

The same procedure and scale equation apply to finding the normal component of the relative acceleration of two points on a link. As Fig. 4-4*c* illustrates, in this case it is the relative velocity of the two points that is laid out at one of them.

4-5. Coriolis Component of Acceleration

The acceleration image method discussed and illustrated in Arts. 4-2 and 4-3 is restricted in application. It applies where a complete analysis can proceed in a step-by-step fashion, from the acceleration of one point on a link to that of a second point which is common to the first and a second link, then to the acceleration of a point common to the second and a third link, and so on. But sometimes this method, utilizing pin joints as turning points in the analysis, is not possible. It may be necessary to go directly from the acceleration of one point on a link to that of a second point on a different link.

In Fig. 4-5*a* body *m* has plane motion. The path *n-n* is fixed on the body. Point *P* is moving along the path, and *Q* is a point fixed on the body which is momentarily coincident with *P*. The center of curvature of the path is *C*, and the radius of curvature is r_{CQ}.

An equation relating the accelerations of points *P* and *Q* can be written by applying Eq. 4-1.

$$\mathbf{a}_P = \mathbf{a}_Q + \mathbf{a}_{P/Q} \tag{4-6}$$

However, Eq. 4-2 no longer applies, for *P* and *Q* are not points on the same link. Instead, the relative acceleration, $\mathbf{a}_{P/Q}$, is given by the following equation:

$$\mathbf{a}_{P/Q} = \mathbf{a}^n_{P/Q} + \mathbf{a}^t_{P/Q} + \mathbf{a}^c \tag{4-7}$$

Substituting for $\mathbf{a}_{P/Q}$ in Eq. 4-6, the following equation results:

$$\mathbf{a}_P = \mathbf{a}_Q + \mathbf{a}^n_{P/Q} + \mathbf{a}^t_{P/Q} + \mathbf{a}^c \tag{4-8}$$

The first of the right-hand terms in Eq. 4-8 is the absolute acceleration of the coincident point, *Q*, which is fixed on the body. The resultant of the second and third terms is the acceleration of the moving point *P* relative to the path. And the last term, $\mathbf{a}^c$, is the *Coriolis component.*

The values of the magnitudes of $\mathbf{a}^n_{P/Q}$ and $\mathbf{a}^t_{P/Q}$ can be found by regarding *P* as revolving instantaneously about the center of curvature, *C*, and applying Eqs. 2-14*a* and 2-15. The resulting equations are

$$a^n_{P/Q} = \frac{V^2_{P/Q}}{r_{CQ}} \tag{4-9}$$

$$a^t_{P/Q} = r_{CQ} \tag{4-10}$$

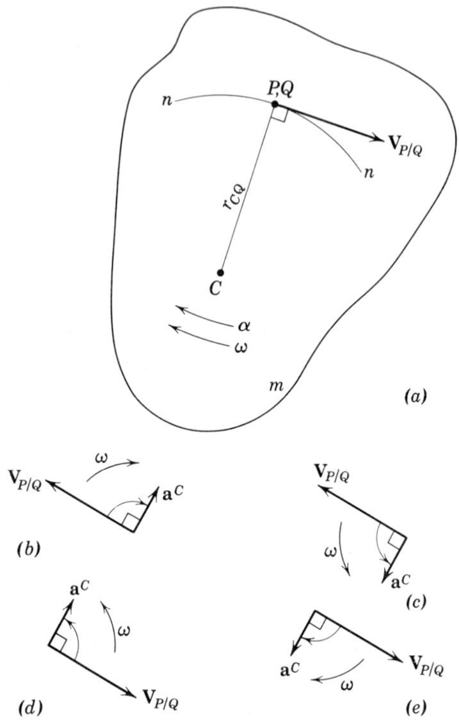

Fig. 4-5. Coriolis component of acceleration.

The magnitude of the Coriolis component is

$$a^c = 2V_{P/Q} \cdot \omega \tag{4-11}$$

In Eqs. 4-8 to 4-11,

P = point moving relative to path
Q = point fixed on body containing path and momentarily coincident with P
C = center of curvature of path
$\mathbf{a}_P$ = absolute acceleration of P
$\mathbf{a}_Q$ = absolute acceleration of Q
$\mathbf{a}^n_{P/Q}$ = normal component of P's acceleration relative to the path
$\mathbf{a}^t_{P/Q}$ = tangential component of P's acceleration relative to the path
$\mathbf{a}^c$ = Coriolis component of acceleration
$V_{P/Q}$ = magnitude of the velocity of P relative to Q (that is, relative to the path)
r_{CQ} = radius of curvature of the path
ω = angular speed of body containing path
α = magnitude of angular acceleration of body containing path

The directions of the last three terms of Eq. 4-8 are given by the following rules:

1. The normal component, $\mathbf{a}^n_{P/Q}$, is normal to the path (along the radius of curvature) and toward the center of curvature.
2. The tangential component, $\mathbf{a}^t_{P/Q}$, is tangent to the path (therefore perpendicular to the radius of curvature).
3. The direction of the Coriolis component, $\mathbf{a}^c$, is the direction the relative velocity vector, $\mathbf{V}_{P/Q}$, would have if rotated about its origin 90° in the direction of the angular velocity of the body containing the path.

For example, in Fig. 4-5*a* the normal component is along the radius and toward *C*. The tangential component is tangent to the path. The Coriolis component is along the radius and toward *C*.

Figures 4-5*b*, *c*, *d*, and *e* illustrate four possible applications of the directional rule for the Coriolis component.

Example 4-3. Determine the velocity and acceleration of the follower body in the cam mechanism of Fig. 4-6*a*.

Solution. The follower body, link 4, has a translatory motion. Therefore it will suffice to find the velocity and acceleration of any point on the body. It will be convenient to use point P_4, the center of the roller, since this is a point common to links 3 and 4. To make the transition in analysis from link 2, whose velocity and acceleration are known, to links 3 and 4, point P_2 on the cam and momentarily coincident with P_4 will be used.

To find the velocity of the follower:

1. The equation relating the velocities of P_2 and P_4 is

$$\mathbf{V}_{P_4} = \mathbf{V}_{P_2} + \mathbf{V}_{P_4/P_2}$$

2. $\mathbf{V}_{P_4}$ is necessarily vertical, for the follower body has a vertical motion. $\mathbf{V}_{P_2}$ is perpendicular to the radius O_2P_2. And the relative velocity, $\mathbf{V}_{P_4/P_2}$, is perpendicular to line CP_4, for, as Fig. 4-6*a* indicates, P_4 moves on a circular path relative to the cam.

3. The magnitude of $\mathbf{V}_{P_2}$ is

$$\begin{aligned} V_{P_2} &= r_{OP}\omega_2 \\ &= (2.78/12)(62.83) \\ &= 14.56 \text{ ft/sec}^2 \end{aligned}$$

4. The completed velocity diagram is shown in Fig. 4-6*b*.

To find the follower acceleration:

1. Points P_2 and P_4 are coincident points on different links. Therefore Eq. 4-8 can be applied.

$$\mathbf{a}_{P_4} = \mathbf{a}_{P_2} + \mathbf{a}^n_{P_4/P_2} + \mathbf{a}^t_{P_4/P_2} + \mathbf{a}^c$$

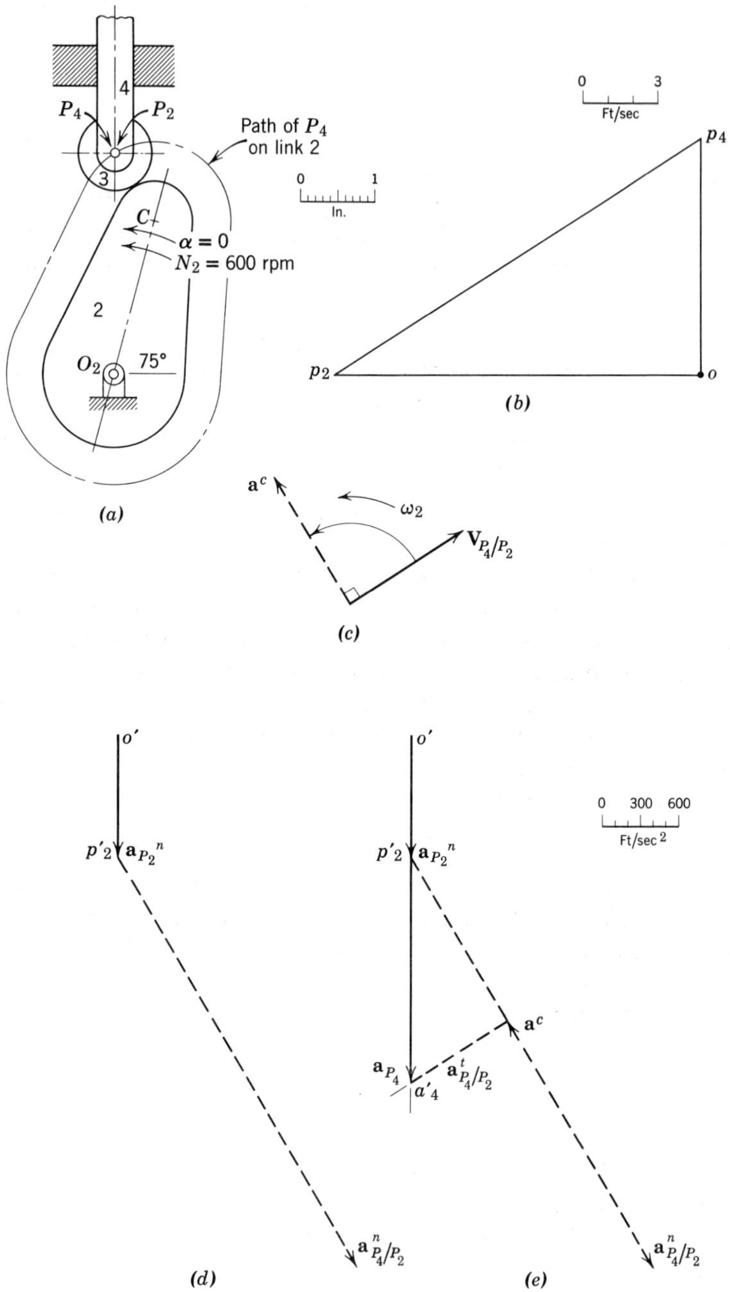

Fig. 4-6. Acceleration analysis of a cam mechanism.

2. And, since point P_2 revolves about a fixed center,

$$\mathbf{a}_{P_2} = \mathbf{a}_{P_2}{}^n + \mathbf{a}_{P_2}{}^t$$

But the cam has no angular acceleration; so

$$\mathbf{a}_{P_2} = \mathbf{a}_{P_2}{}^n$$

and

$$\mathbf{a}_{P_4} = \mathbf{a}_{P_2}{}^n + \mathbf{a}^n_{P_4/P_2} + \mathbf{a}^t_{P_4/P_2} + \mathbf{a}^c$$

3. The magnitudes of the normal components and the Coriolis component can be calculated.

$$\begin{aligned} a_{P_2}{}^n &= \frac{V_{P_2}{}^2}{r_{OP}} \\ &= \frac{(14.55)^2}{(2.78/12)} \\ &= 914 \text{ ft/sec}^2 \\ a^n_{P_4/P_2} &= \frac{V^2_{P_4/P_2}}{r_{CP}} \\ &= \frac{(17)^2}{(0.98/12)} \\ &= 3540 \text{ ft/sec}^2 \\ a^c &= 2V_{P_4/P_2}\omega_2 \\ &= \frac{(2)(17)(2\pi)(600)}{60} \\ &= 2135 \text{ ft/sec}^2 \end{aligned}$$

4. Next the directions of the acceleration vectors should be determined. Since the follower motion is vertical, the acceleration of the follower, $\mathbf{a}_{P_4}$, is necessarily vertical. The acceleration conponent $\mathbf{a}_{P_2}{}^n$ is toward O_2, the center of rotation of the cam. The normal component of the relative acceleration, $\mathbf{a}^n_{P_4/P_2}$, is toward C, and the tangential component, $\mathbf{a}^t_{P_4/P_2}$, is perpendicular to CP_4. The Coriolis component, $\mathbf{a}^c$, has the direction the relative velocity vector, $\mathbf{V}_{P_4/P_2}$, would have if rotated about its origin for 90° in the direction of the cam's velocity. This rotation is illustrated (not to scale) in Fig. 4-6*c*.

5. The acceleration diagram can now be constructed. The strategy at this point is the same as that used in the previous examples. Those accelerations whose magnitudes are unknown are drawn last, their

intersection completing the diagram. The following steps were used in constructing the acceleration diagram:

(*a*) From pole point o', $\mathbf{a}_{P_2}{}^n$ was drawn (Fig. 4-6*d*).
(*b*) From the terminus of $\mathbf{a}_{P_2}{}^n$, the acceleration $\mathbf{a}^n_{P_4/P_2}$ was drawn (Fig. 4-6*d*).
(*c*) From the terminus of $\mathbf{a}^n_{P_2/P_2}$, the Coriolis component, $\mathbf{a}^c$, was drawn (Fig. 4-6*e*). Note the overlap.
(*d*) From the terminus of the Coriolis component a line was drawn of indefinite length to represent the tangential component, $\mathbf{a}^t_{P_4/P_2}$.
(*e*) From o' a line was drawn in the direction of $\mathbf{a}_{P_4}$, that is, vertically.
(*f*) The intersection of $\mathbf{a}^t_{P/_4P_2}$ and $\mathbf{a}_{P_4}$ is p'_4. The diagram is complete.

$\mathbf{a}_{P_4} = \overrightarrow{o'a'}_4$.

Example 4-4. Find the acceleration of the ram, link 6, of the quick-return mechanism in Fig. 4-7*a*. The crank turns at a constant velocity, and the speed of point A_2 is 20 ft/sec, where A_2 is the center of the pin joint connecting the crank and the slider.

Solution. To find the acceleration of C, a point on the ram, first the acceleration of A_4, and then that of B, will have to be found. And, of course, before these accelerations can be determined, the velocity diagram must be constructed.

To construct the velocity diagram:

1. In the analysis, points A_2 and A_4 will be used. A_2 is the center of the pin joint connecting links 2 and 3, while A_4 is on link 4 and momentarily coincident with A_2.

2. The velocity equation relating points A_2 and A_4 is

$$\mathbf{V}_{A_4} = \mathbf{V}_{A_2} + \mathbf{V}_{A_4/A_2}$$

where $\mathbf{V}_{A_4}$ is perpendicular to line O_4A_4 and $\mathbf{V}_{A_4/A_2}$ is parallel to link 4.

3. In Fig. 4-7*b*, $\mathbf{V}_{A_2}$ was laid out from o to scale. From its terminus was drawn a line of indefinite length in the direction of $\mathbf{V}_{A_4/A_2}$. And from o a line of indefinite length was drawn in the direction of $\mathbf{V}_{A_4}$. The intersection point is a_4.

4. To determine the velocity of point B, the image principle was used; for B is collinear with O_4 and A_4. In the diagram, b was located by determining distance ob, using the proportion:

$$\frac{ob}{oa_4} = \frac{O_4B}{O_4A_4}$$

5. The velocity of C was found by solving the vector equation

$$\mathbf{V}_C = \mathbf{V}_B + \mathbf{V}_{C/B}$$

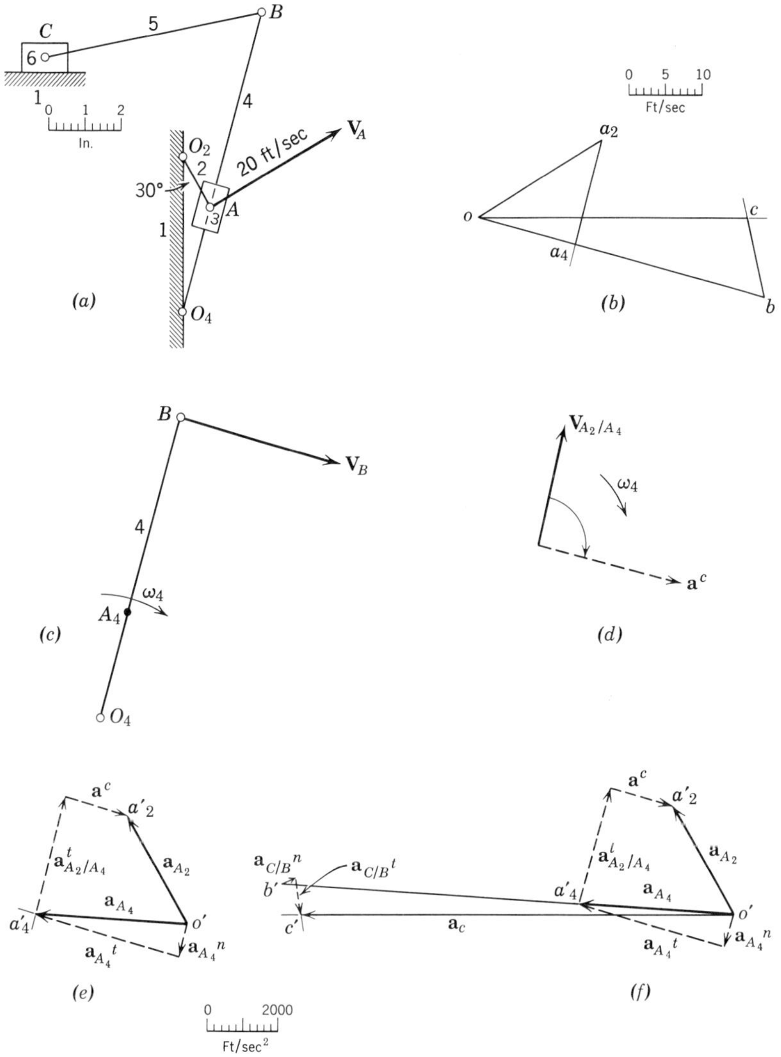

Fig. 4-7. Acceleration analysis of a quick-return mechanism.

$\mathbf{V}_C$ is necessarily horizontal, $\mathbf{V}_B$ is known, and $\mathbf{V}_{C/B}$ is perpendicular to line *CB*.

6. In Fig. 4-7*b* the velocity diagram was completed by drawing two lines of indefinite length, one from *o* in the direction of $\mathbf{V}_C$ and the other from *b* in the direction of $\mathbf{V}_{C/B}$. Their intersection is *c*.

To find $\mathbf{a}_{A_4}$:

1. Before writing an acceleration equation relating points A_2 and A_4, a choice must be made. The equation might be written regarding A_2 as moving relative to link 4, or by regarding point A_4 as moving relative to link 2. The path of A_2 relative to link 4 is a straight line, while that of A_4 relative to link 2 is a complex curve. The simpler alternative is to write the equation for the motion of A_2 relative to link 4. The equation is

$$\mathbf{a}_{A_2} = \mathbf{a}_{A_4} + \mathbf{a}^n_{A_2/A_4} + \mathbf{a}^t_{A_2/A_4} + \mathbf{a}^c$$

Solving for $\mathbf{a}_{A_4}$,

$$\mathbf{a}_{A_4} = \mathbf{a}_{A_2} - \mathbf{a}^n_{A_2/A_4} - \mathbf{a}^t_{A_2/A_4} - \mathbf{a}^c$$

Since A_2 is on a link rotating about a fixed center,

$$\mathbf{a}_{A_2} = \mathbf{a}_{A_2}{}^n + \mathbf{a}_{A_2}{}^t$$

But link 2 turns at a constant angular velocity; so

$$\mathbf{a}_{A_2} = \mathbf{a}_{A_2}{}^n$$

The movement of A_2 relative to link 4 is on a straight line. Therefore the normal component $\mathbf{a}^n_{A_2/A_4}$ is zero. (In Eq. 4-9, if r_{CQ} becomes infinitely large, $a^n_{P/Q}$ becomes zero.)

The equation for $\mathbf{a}_{A_4}$ can now be modified to read

$$\mathbf{a}_{A_4} = \mathbf{a}_{A_2}{}^n - \mathbf{a}^t_{A_2/A_4} - \mathbf{a}^c$$

But further, since A_4 is on a link turning about a fixed center,

$$\mathbf{a}_{A_4} = \mathbf{a}_{A_4}{}^n + \mathbf{a}_{A_4}{}^t$$

A simultaneous solution of these last two vector equations must be made.

2. The magnitudes of the normal components and the Coriolis component are

$$\begin{aligned} a_{A_2}{}^n &= \frac{V_{A_2}{}^2}{r_{A_2}} \\ &= \frac{(20)^2}{(1.5/12)} \\ &= 3200 \text{ ft/sec}^2 \end{aligned}$$

$$\begin{aligned} a^c &= 2V_{A_2/A_4}\omega_4 \\ &= 2V_{A_2/A_4}\frac{V_B}{r_B} \\ &= \frac{(2)(14.3)(40)}{(8/12)} \\ &= 1716 \text{ ft/sec}^2 \end{aligned} \qquad \begin{aligned} a_{A_4}{}^n &= \frac{V_{A_4}{}^2}{r_{A_4}} \\ &= \frac{(14)^2}{(2.75/12)} \\ &= 856 \text{ ft/sec}^2 \end{aligned}$$

3. The direction of $\mathbf{a}_{A_2}{}^n$ must be from A_2 toward O_2. That of $\mathbf{a}^t_{A_2/A_4}$ must be parallel to O_4B. The sketch of link 4 in Fig. 4-7*c*, with $\mathbf{V}_B$ indicated, shows that link 4 is turning clockwise. Therefore the direction of $\mathbf{a}^c$ is the direction $\mathbf{V}_{A_2/A_4}$ would have if rotated about its origin 90° clockwise, as shown in Fig. 4-7*d*. The direction of $\mathbf{a}_{A_4}{}^n$ is from A_4 toward O_4.

4. The construction of that portion of the acceleration diagram necessary to determine $\mathbf{a}_{A_4}$ is shown in Fig. 4-7*e*. The following procedure was used:

(*a*) From o' the acceleration $\mathbf{a}_{A_2} = \mathbf{a}_{A_2}{}^n$ was laid off.
(*b*) Then $\mathbf{a}^c$ was *subtracted* from $\mathbf{a}_{A_2}$ by drawing it with its terminus at that of $\mathbf{a}_{A_2}$.
(*c*) Next, from the origin of $\mathbf{a}^c$ was drawn a line of indefinite length in the direction of $\mathbf{a}^t_{A_2/A_4}$.
(*d*) From o' $\mathbf{a}_{A_4}{}^n$ was laid out.
(*e*) From its terminus a line of indefinite length was drawn in the direction of $\mathbf{a}_{A_4}{}^t$.
(*f*) The intersection point is a'_4, and the vector from o' to a'_4 represents $\mathbf{a}_{A_4}$.
(*g*) The sense of $\mathbf{a}^t_{A_2/A_4}$ is upward, for it too was being subtracted from $\mathbf{a}_{A_2}$.

To find $\mathbf{a}_B$:

1. Point B is collinear with points O_4 and A_4, which are also on link 4. The image principle should be applied.

2. Point b' is located in the acceleration diagram by use of the proportion:

$$\frac{o'b'}{o'a'_4} = \frac{O_4B}{O_4A_4}$$

To find $\mathbf{a}_C$:

1. Since points C and B are on the same link,

$$\mathbf{a}_C = \mathbf{a}_B + \mathbf{a}^n_{C/B} + \mathbf{a}^t_{C/B}$$

2. The normal component $\mathbf{a}^n_{C/B}$ has a sense from C toward B, and its magnitude is:

$$a^n_{C/B} = \frac{V^2_{C/B}}{r_{CB}} = \frac{(10.9)^2}{(6/12)} = 237 \text{ ft/sec}^2$$

3. Acceleration $\mathbf{a}_C$ is necessarily horizontal, and $\mathbf{a}_B$ is already known. Tangential component, $\mathbf{a}^t_{C/B}$, is perpendicular to line CB.
4. The acceleration diagram has been completed in Fig. 4-7*f*.

   (*a*) From the terminus of $\mathbf{a}_B$, the normal component $\mathbf{a}^n_{C/B}$ was laid out.
   (*b*) From the terminus of $\mathbf{a}^n_{C/B}$, a line of indefinite length was drawn in the direction of $\mathbf{a}^t_{C/B}$.
   (*c*) From o' a line of indefinite length was drawn in a horizontal direction, the direction of the ram's movement.
   (*d*) The intersection point is c', and the vector from o' to c' represents $\mathbf{a}_C$.

4-6. Equivalent Mechanisms

Acceleration analyses of mechanisms that involve motion transmission by direct contact, such as cam mechanisms, can be greatly facilitated if the actual mechanism is replaced by a simpler, but kinematically equivalent, one for the instant at which the analysis is being made.

In devising an equivalent mechanism to replace a direct contact mechanism, the centers of curvature of the link profiles at the point of contact must be known. In most cases they will be known, but occasionally their locations will have to be determined by a graphical method.

A graphical technique for the location of the center of curvature is illustrated in Fig. 4-8. A mirror or strip of reflecting steel is used. The reflecting surface is placed perpendicular to the paper and through the point on the curve at which the center of curvature is to be located.

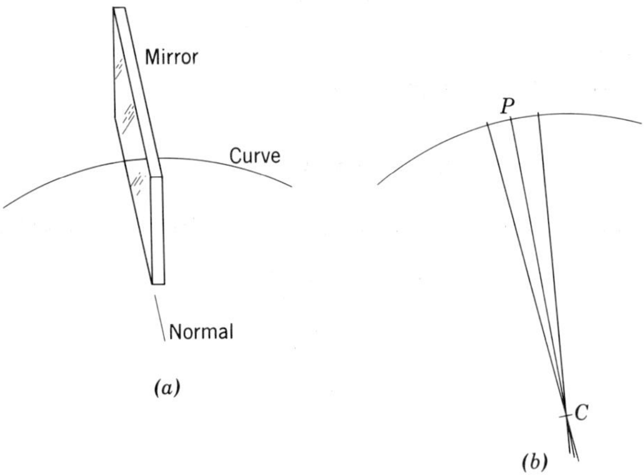

Fig. 4-8. Graphical determination of center of curvature.

The mirror is next pivoted about this point until the drawn portion of the curve and the reflected image form a smooth curve. The edge of the mirror is then along the common normal to the curve at the point. By repeating this process at a point a short distance along the curve from the given point, a pair of normals are determined which intersect at the center of curvature. As a check, the process should be repeated for a third point near the given point, but on the other side, as has been done in Fig. 4-8*b*.

The replacement of a direct contact mechanism by a four-bar mechanism is shown in Fig. 4-9. In this instance there are two curved surfaces in

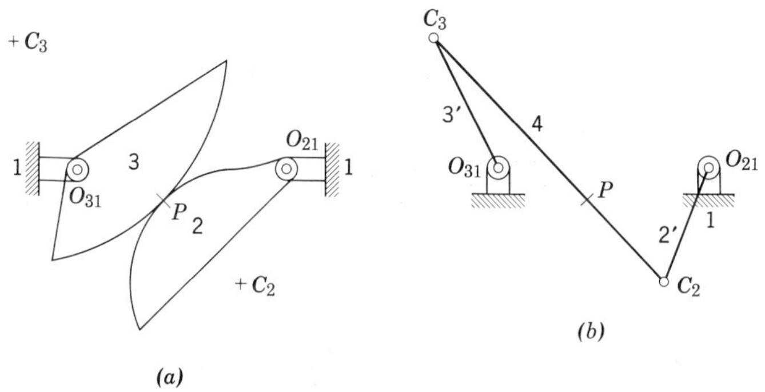

Fig. 4-9. Equivalent mechanism for direct contact mechanism.

direct contact at P. As a first step the centers of curvature are located. They are C_2 and C_3. The mechanism can then be replaced by a four-bar mechanism. The original link 2 is replaced by link 2′, which connects the center of rotation, O_2, and the centre of curvature, C_2. Similarly, link 3 is replaced by link 3′, which connects O_3 and C_3. The coupler of the four-bar mechanism, link 4, connects the two centers of curvature.

Further illustrations of the replacement of a mechanism by a four-bar mechanism are given in Figs. 4-10 and 4-11. The use of the first inversion of the sliding-block chain (where the guide is the fixed link) is illustrated in Figs. 4-12 and 4-13. And in Fig. 4-14 a Scotch yoke has become the equivalent mechanism. A different inversion of the sliding-block chain has been used in Fig. 4-15.

In each case, it must be remembered, the equivalent mechanism is, in fact, only equivalent at the instant. Later, both the configuration and the proportions of the equivalent mechanism will have changed.

A good catalog of equivalent mechanisms for cam mechanisms is given in an article by Rothbart (Ref. 3). Special techniques for acceleration analysis in difficult cases are discussed in Refs. 1 and 2.

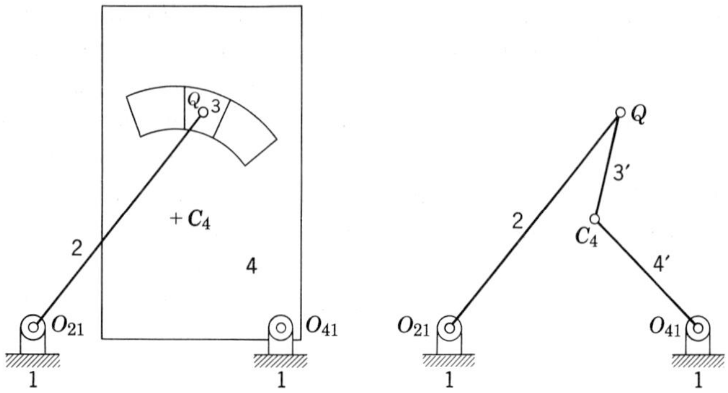

(a) (b)

Fig. 4-10. Equivalent mechanism.

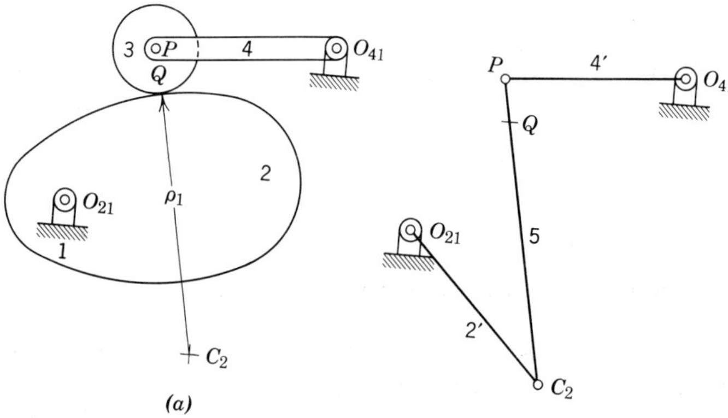

(a) (b)

Fig. 4-11. Equivalent mechanism for disk cam with oscillating roller-face follower.

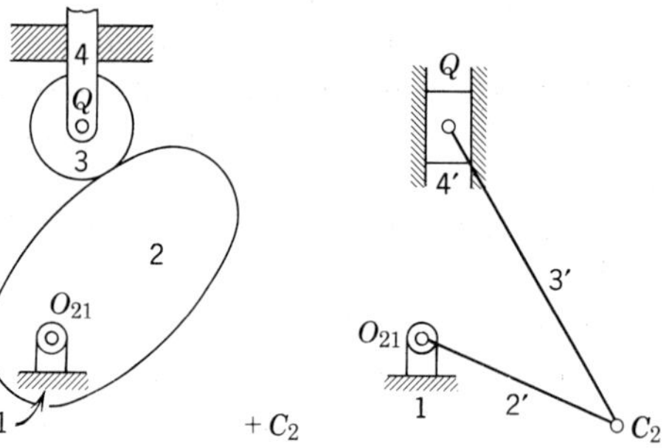

(a) (b)

Fig. 4-12. Equivalent mechanism for disk cam with reciprocating roller-face follower.

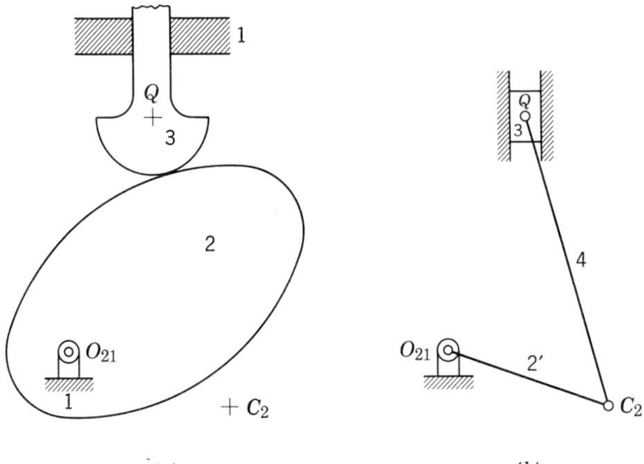

(a) (b)

Fig. 4-13. Equivalent mechanism for disk cam with reciprocating curved-face follower.

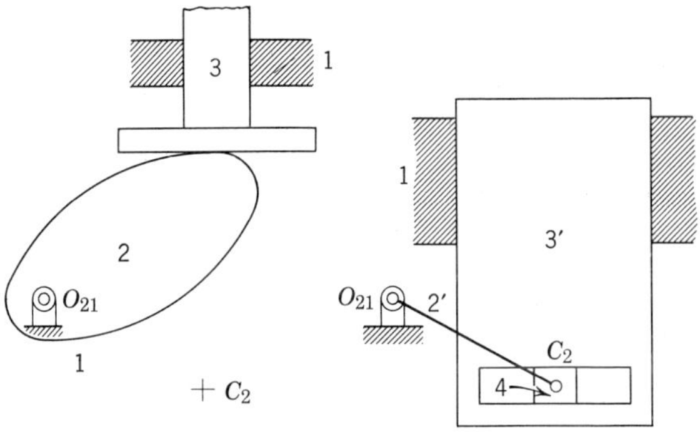

(a) (b)

Fig. 4-14. Equivalent mechanism for disk cam with reciprocating flat-face follower.

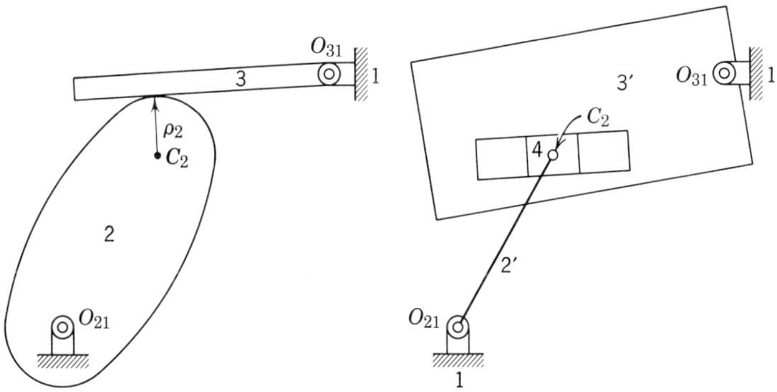

(a) (b)

Fig. 4-15. Equivalent mechanism for disk cam with oscillating flat-face follower.

REFERENCES

1. A. S. Hall and E. S. Ault, "Auxiliary Points Aid Acceleration Analysis," *Machine Design*, November 1943.
2. Rolland T. Hinkle, "Analyzing Accelerations of Complex Mechanisms," *Machine Design*, May 1952.
3. Harold A. Rothbart, "Equivalent Mechanisms for Cams," *Machine Design*, March 20, 1958.

PROBLEMS

Unless specified otherwise, the angular velocity of the driving link in each mechanism is constant.

4-1. Construct the acceleration diagram for the mechanism of Fig. P3-2 and determine the angular acceleration of link 3.

4-2. Construct the acceleration diagram for the mechanism of Fig. P3-3 and determine the angular accelerations of links 3 and 4.

4-3. Construct the acceleration diagram for the mechanism of Fig. P3-4 and determine the angular accelerations of links 3, 4, and 5.

4-4. Construct the acceleration diagram for the mechanism of Fig. P3-5 and determine the angular accelerations of links 3 and 5.

4-5. Construct the acceleration diagram for the mechanism in Fig. P3-7 and determine the angular accelerations of links 3, 4, and 5.

4-6. Construct the acceleration diagram for the mechanism of Fig. P3-8 and determine the angular accelerations of links 3, 4, and 5.

4-7. Construct the acceleration diagram for the mechanism of Fig. P3-9 and determine the angular accelerations of links 3, 4, and 5.

4-8. Construct the acceleration diagram for the mechanism of Fig. P3-10 and determine the angular accelerations of links 3 and 5.

4-9. Determine the acceleration of link 3 in the mechanism of Fig. P3-1.

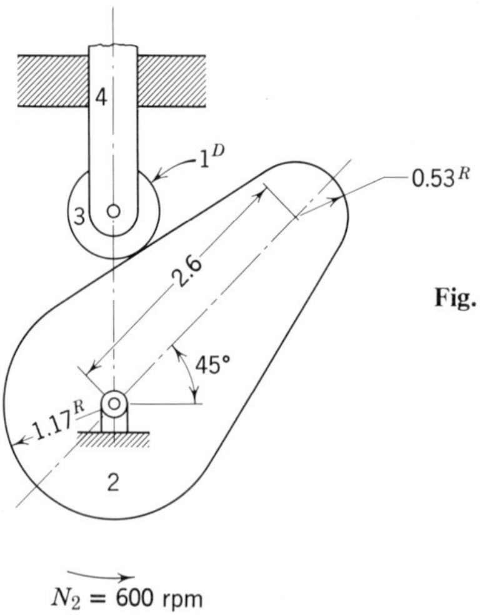

Fig. P4-11

4-10. Construct the acceleration diagram for the mechanism of Fig. P3-6 and determine the angular accelerations of links 4 and 5.

4-11. The cam in Fig. P4-11 is turning at a constant angular velocity of 600 rpm counterclockwise. Determine the acceleration of the follower body, link 4.

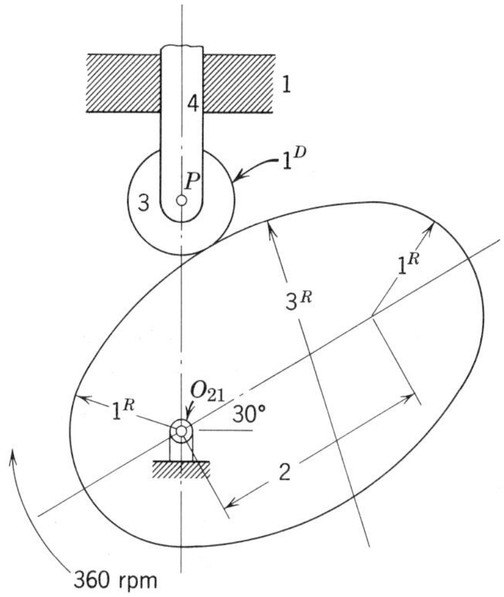

Fig. P4-12

4-12. The cam in Fig. P4-12 is turning at a constant angular velocity of 360 rpm clockwise. Determine the acceleration of the follower body, link 4.

5 Linkages

5-1. Introduction

Modern demands for dependable and accurate means for control and computation have aroused intense interest in the possibilities of mechanisms derived from simple kinematic chains. Particular interest has been expressed, and impressive research done, in the simplest of these, the four-bar chain and the sliding-block chain.

Reliance on cut-and-try procedures, which puts maximum emphasis on the ingenuity of the individual designer, is giving way to a search for a more fundamental understanding of these devices. Systematic design techniques of great power and wide application are being developed which will permit the custom tailoring of linkages to meet the multitude of demands made by modern technology, demands for devices that permit accurate control of processes and machines or that relieve men of the drudgery of routine and repetitive calculations.

Although this chapter will be restricted to the more elementary aspects of the analysis and design of linkages, it is hoped that the student will glimpse the enormous and barely tapped possibilities of these simple mechanisms.

5-2. Four-Bar Mechanisms

The most fundamental of the plane kinematic chains is the four-bar chain, shown in Fig. 5-1. It consists of four rigid links connected in the form of a quadrilateral by four pin joints.

When a four-bar mechanism is made from a four-bar chain by fixing one link, the fixed link is called the *frame* or the *line of centers*. Pinned to the frame are two links, each of which is either a crank or a lever. If it can rotate continuously, it is a *crank*. It is a *lever* if restricted to an oscillatory motion. At times it is not yet known whether one of these two links is in fact a crank or a lever. In the event of this uncertainty, it is customary to

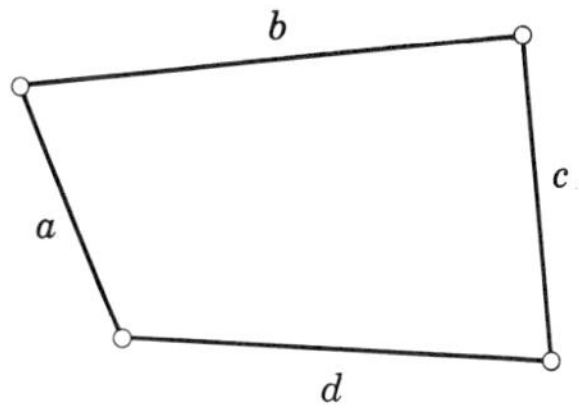

Fig. 5-1. The four-bar chain.

refer to it temporarily as a crank. Joining (or coupling) the cranks is the *coupler*, the link opposite the frame.

Classification of four-bar chains. In any four-bar chain, let the length of the shortest link be a and the lengths of the succeeding links be b, c, and d, reading in order, clockwise or counterclockwise, around the chain. Four-bar chains can be divided into two classes:

Class 1. A Class 1 four-bar chain satisfies both of the criteria:

$$d < (b + c - a) \tag{5-1}$$

$$d > (|b - c| + a) \tag{5-2}$$

These two criteria can, when convenient, be combined into a single statement:

$$(|b - c| + a) < d < (b + c - a) \tag{5-3}$$

Class 2. Any four-bar chain which is physically possible but fails to satisfy both of the Class 1 criteria is a Class 2 chain.

Mechanisms made from a Class 1 four-bar chain can accept or produce a continuous rotary motion, for the shortest link can rotate continuously relative to the others. This is not the case for mechanisms derived from the Class 2 chain. The links can only oscillate relative to one another.

A four-bar chain is physically impossible if one of the links has a length greater than the sum of the other three. Such a combination is shown in Fig. 5-2. Here link d has a length greater than the sum of a, b, and c. It cannot be assembled.

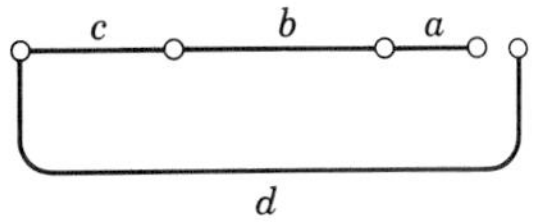

Fig. 5-2. Physically impossible four-bar chain.

Inversions. Designers are frequently interested in the relative motions of the cranks or levers in a four-bar mechanism. From this viewpoint, three types of four-bar mechanism can be distinguished: the crank and rocker mechanism, the drag-link mechanism, and the double-rocker mechanism. Table 5-1 shows how these various types of four-bar mechanism are derived from the four-bar chain.

TABLE 5-1

Class of Chain	Criterion	Mechanism Type
1	Shortest link a crank	Crank and rocker
1	Shortest link the frame	Drag-link
1	Shortest link the coupler	Double-rocker
2	All inversions	Double-rocker

It also shows that crank and rocker or drag-link mechanisms can be made only from a Class 1 four-bar chain, while double-rocker mechanisms can be made from either chain.

In a crank and rocker mechanism one of the links pinned to the frame is a crank while the other is a lever (or rocker). In typical applications a rotary input motion produces an oscillatory output. Two inversions of the Class 1 chain are crank and rocker mechanisms (Figs. 5-3*a*, *b*). In each case the shortest link is a crank.

Both of the links pinned to the frame in a drag-link mechanism are cranks (Fig. 5-3*d*). In typical applications of this mechanism, the driving crank turns at a constant velocity while the driven crank rotates with a varying velocity.

The double-rocker mechanism (Fig. 5-3*c*) has two levers. It could be used with an oscillatory input to generate a different oscillatory output. While double-rocker mechanisms can be derived either from Class 1 or Class 2 chains, there is a fundamental distinction between them. The coupler of the Class 1 double-rocker mechanism can rotate continuously, but this is not possible in a Class 2 double-rocker mechanism. The possibility of continuous rotation of the coupler can be made use of. A flexible shaft connected to the coupler and causing it to rotate would produce two different oscillating outputs. In one type of office fan the motor housing is one lever of a Class 1 double-rocker mechanism. The coupler is pinned to the rear of the motor housing and driven by gearing from the fan motor. A link from the other end of the coupler to the fan base is the second lever. The rotation of the coupler causes the fan housing to oscillate.

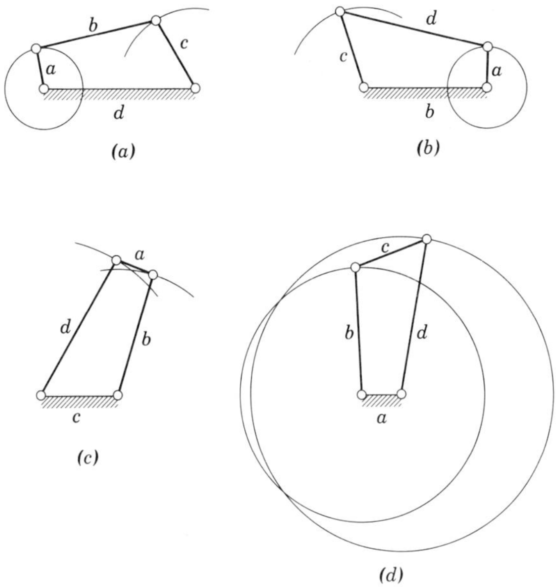

Fig. 5-3. Inversions of a Class 1 four-bar chain ($a = 1, b = 3.2, c = 2, d = 4$). (*a*) Crank and rocker mechanism. (*b*) Crank and rocker mechanism. (*c*) Double-rocker mechanism. (*d*) Drag-link mechanism.

Example 5-1. A crank and rocker mechanism is to be made. The crank length is to be 1.0 in., the coupler length 2.9 in., and the lever length 1.6 in. What would be a suitable frame length?

Solution. A crank and rocker mechanism is made from a Class 1 four-bar chain by making the shortest link a crank. Thus $a = 1.0$ in., $b = 2.9$ in., and $c = 1.6$ in. The length of the frame is d. Applying the two criteria for a Class 1 four-bar chain,

$$d < (b + c - a)$$

$$d < (2.9 + 1.6 - 1.0)$$

$$d < 3.5$$

$$d > (|b - c| + a)$$

$$d > (|2.9 - 1.6| + 1.0)$$

$$d > 2.3$$

The frame length must be greater than 2.3 in. but less than 3.5 in.

5-3. Angular Velocity Ratios

Instant center analysis. The ratio of the angular velocities of the input and output links of a four-bar mechanism can be found by applying the instant center techniques. A crank and rocker mechanism is shown in Fig. 5-4 at two different phases. Link 2 is the driver and link 4 the driven.

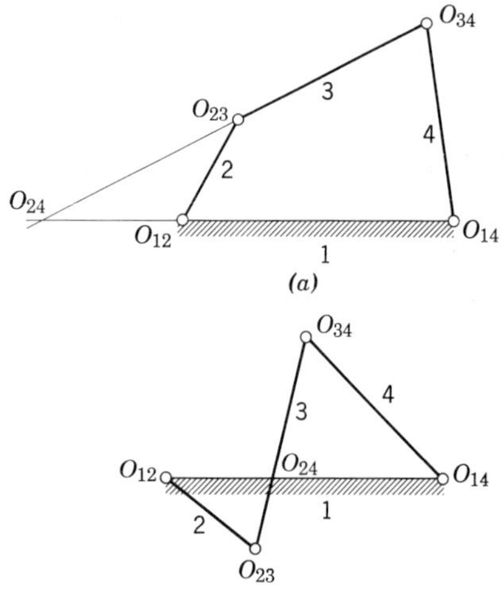

(b)

Fig. 5-4. Angular velocity ratio by instant centers.

It is required to find the ratio of the angular velocities of the driven to driver. If the instant center methods of Art. 3-10 are used, it can be stated that the speeds of links 2 and 4 vary inversely as the distances from their common instant center to the fixed centers. That is,

$$\frac{\omega_4}{\omega_2} = \frac{O_{12}O_{24}}{O_{14}O_{24}} \tag{5-4}$$

If the common instant center lies between the fixed centers, the links turn in opposite directions; otherwise they turn in the same direction. For the mechanism of Fig. 5-4*a*, the common instant center, O_{24}, lies outside the fixed centers (O_{12} and O_{14}). Thus at the instant shown links 2 and 4 are turning in the same direction. In Fig. 5-4*b*, on the other hand, the common instant center lies between the fixed centers. In this case links 2 and 4 turn in opposite directions.

For a four-bar mechanism, the common instant center for the two cranks (or levers) can be found quickly without explicit use of the Aronhold-Kennedy Theorem. As the examples in Fig. 5-4 indicate, the common instant center is always at the intersection of the coupler and frame, each extended if necessary.

Example 5-2. A crank and rocker mechanism is shown at 105° phase in Fig. 5-5*a*. The instantaneous angular velocity of the driver, link 2, is 12.4 rad/sec clockwise. Determine the instantaneous angular velocity of link 4.

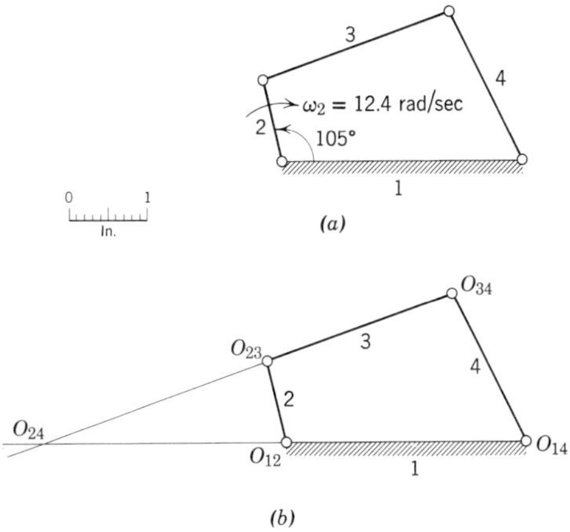

Fig. 5-5. Example problems 5-2, 5-3.

Solution. The first step is to locate the common instant center, O_{24}, and to label the fixed centers, O_{12} and O_{14}. This has been done in Fig. 5-5*b*. First, the fixed centers were located by inspection and labeled. Then the common instant center, O_{24}, was located at the intersection of the coupler and the frame, each extended. By measurement on the original drawing, $O_{12}O_{24} = 3.20$ and $O_{14}O_{24} = 6.21$. If Eq. 5-4 is applied, the speed of link 4 can be found.

$$\frac{\omega_4}{\omega_2} = \frac{O_{12}O_{24}}{O_{14}O_{24}} \tag{5-4}$$

$$\frac{\omega_4}{12.4} = \frac{3.20}{6.21}$$

$$\omega_4 = 6.39 \text{ rad/sec}$$

Since common instant center, O_{24}, falls outside the fixed centers, links 2 and 4 turn in the same direction. Therefore the velocity of link 4 at the instant is 6.39 rad/sec clockwise.

Mechanical advantage. The velocity analysis of a mechanism furnishes information that can be used directly in computing the mechanism's mechanical advantage at the instant. The technique will be illustrated by discussion of the mechanical advantage of a four-bar mechanism. In this illustration it is assumed that one crank is the driver while the other is driven.

If a torque is applied to (or by) a rotating body, the power is

$$P = T\omega \tag{5-5}$$

where P = power, in.-lb/sec
T = torque, in.-lb
ω = angular speed, rad/sec
Let P_i = input power
P_o = output power
T_i = input torque
T_o = output torque
ω_i = input speed
ω_o = output speed

Then, if friction is neglected,

$$P_i = P_o$$

$$T_i\omega_i = T_o\omega_o$$

The ratio of the output torque to the input torque is the mechanism's mechanical advantage (MA). Therefore

$$\text{MA} = \frac{T_o}{T_i} = \frac{\omega_i}{\omega_o} \tag{5-6}$$

The mechanical advantage is equal to the ratio of the input speed to the output speed.

Example 5-3. Find the mechanical advantage of the four-bar mechanism of Fig. 5-5 for the instant shown.

Solution. Link 2 is the input link and link 4 the output. Applying Eq. 5-6,

$$\text{MA} = \frac{\omega_2}{\omega_4}$$

Then, from Eq. 5-4,

$$\text{MA} = \frac{\omega_2}{\omega_4} = \frac{O_{14}O_{24}}{O_{12}O_{24}}$$

Substituting the values of $O_{14}O_{24}$ and $O_{12}O_{24}$,

$$\text{MA} = \frac{6.21}{3.20} = 1.94$$

Dead points. If at some phase the mechanical advantage of a mechanism becomes zero, that phase is a *dead point*. This is illustrated in Fig. 5-6. A four-bar mechanism is shown at a dead point. Since link 2 is the driver and link 4 the driven,

$$\text{MA} = \frac{O_{14}O_{24}}{O_{12}O_{24}}$$

But at this phase the coupler produced intersects the frame at O_{14}. Thus O_{14} and O_{24} are coincident, and $O_{14}O_{24} = 0$. This, in turn, implies that

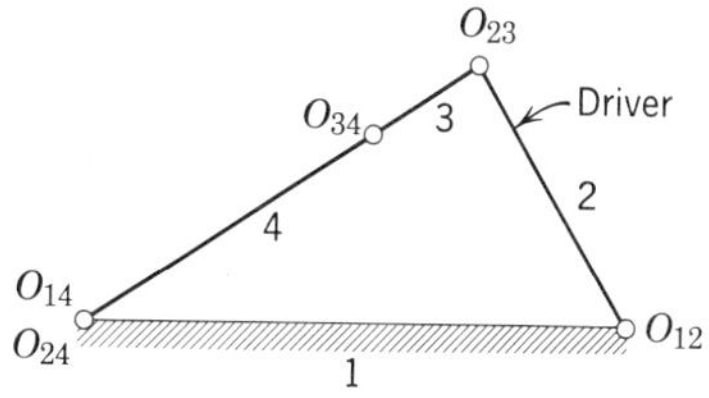

Fig. 5-6. Dead points.

MA = 0. The driver by itself cannot move the mechanism past a dead point. In practice, dead points must be avoided or external means provided to carry the mechanism past a dead point.

5-4. Virtual Centers and Relative Position

The notions of virtual center (or pole), relative position, and relative pole, have been made the basis of a variety of valuable graphical techniques for the synthesis of mechanisms by the German kinematicians. In this section these ideas will be introduced; in later sections they will be applied in some of the simpler problems of mechanism synthesis.

Virtual pole. Figure 5-7*a* shows a plane *ABC*, which has moved from an initial position $A_1B_1C_1$ to the position $A_2B_2C_2$. For purposes of analysis, the actual plane motion of plane *ABC* can be replaced by a simple rotation about some center, the *virtual center* or *pole*. It is not important that the motion of *ABC* actually was a simple rotation, merely that it could have been.

In any rotation of a body, each point on the body describes a circular arc whose center is the center of rotation. Therefore, if two positions of a point are known, the line joining them must be a chord of the circle; and the midnormal (perpendicular bisector) of this chord must pass through the center of rotation, the virtual center. As a convenient notation the midnormal of chord A_1A_2 will be called 'a_{12}', the midnormal of B_1B_2 will be called 'b_{12}', etc.

The construction for the virtual center, P_{12}, is shown in Fig. 5-7*b*. P_{12} is located at the intersection of midnormals a_{12} (midnormal of A_1A_2) and b_{12}

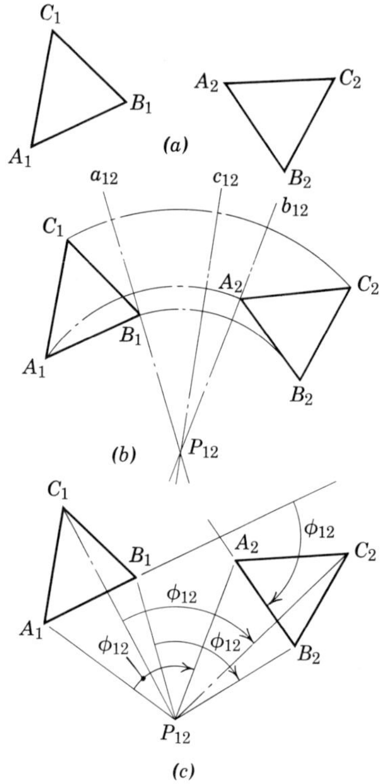

Fig. 5-7. Virtual center or pole.

(midnormal of B_1B_2). As a check, midnormal c_{12} has been drawn. It, of course, should also pass through P_{12}.

The angle of rotation of plane *ABC* is ϕ_{12}. A line joining any point on *ABC* to the pole P_{12} will rotate through this angle, as Fig. 5-7*c* demonstrates. Indeed, any line on the plane, such as line *AB*, will rotate through this same angle.

Relative position. In Fig. 5-8*a* a line *AB* has moved from A_1B_1 to A_2B_2 while a second line has moved from P_1Q_1 to P_2Q_2. It is required to find the displacement of *PQ relative to AB*, that is, as seen by an observer on line *AB*.

The construction is shown in Fig. 5-8*b*. Since it is the movement of PQ relative to AB that is wanted, AB is regarded as fixed in some convenient position, in this case A_1B_1. Relative to A_1B_1, then, P_1Q_1 is the initial

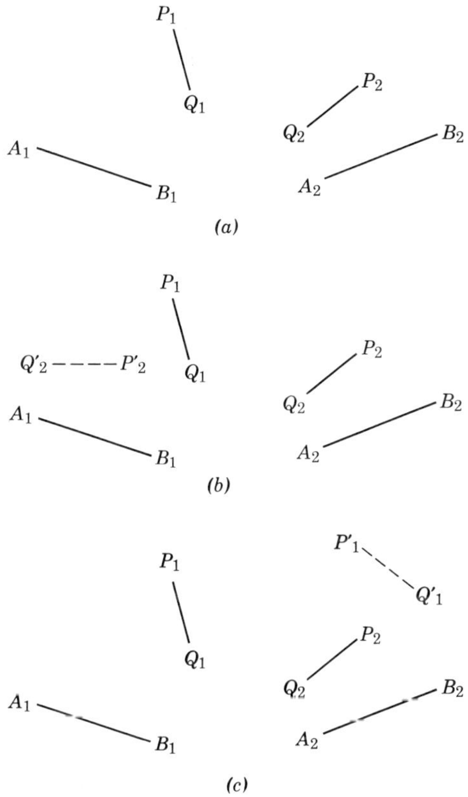

Fig. 5-8. Relative position.

position of PQ. It remains to find the final position of PQ, $P'_2Q'_2$, relative to A_1B_1.

In the final position of PQ, P is a distance P_2A_2 from A and P_2B_2 from B. Similarly, Q is at a distance Q_2A_2 from A and Q_2B_2 from B. $P'_2Q'_2$ can be found either by triangulation or by use of an overlay. To use the triangulation method:

1. Swing arcs of radii A_2P_2 and B_2P_2 about A_1 and B_1 as centers. Their intersection is P'_2.
2. Swing arcs of radii A_2Q_2 and B_2Q_2 about A_1 and B_1 as centers. Their intersection is Q'_2.

For the overlay method:

1. Place a piece of vellum over the drawing and trace A_2B_2 and P_2Q_2.
2. Move the overlay so that A_2B_2 on the overlay coincides with A_1B_1 on the original drawing.
3. Using a pricker, prick through the overlay at P_2 and Q_2 on the overlay to locate points P'_2 and Q'_2 on the original drawing.

Overlay methods for this type of construction are usually the more economical of effort.

In Fig. 5-8*c* the same construction has been made, but this time regarding AB as fixed at A_2B_2. In this case P_2Q_2 is the final position of PQ relative to

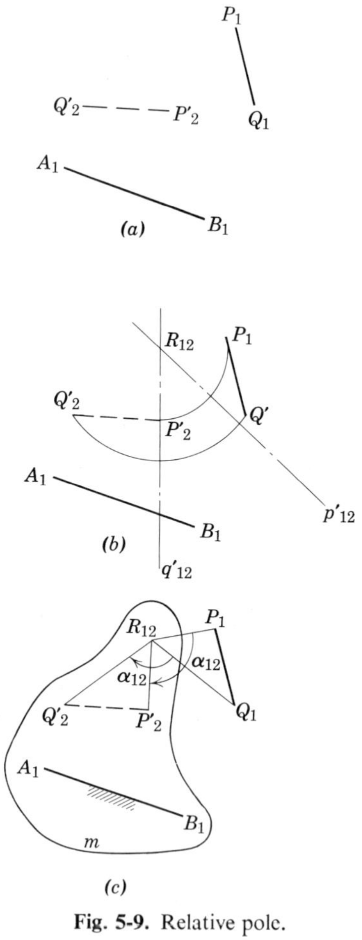

Fig. 5-9. Relative pole.

A_2B_2. It remains to find $P'_1Q'_1$, the initial position of PQ relative to A_2B_2. The overlay method for this construction follows the same pattern as before:

1. On a vellum overlay trace A_1B_1 and P_1Q_1.
2. Move the overlay so that A_1B_1 on the overlay coincides with A_2B_2 on the drawing.
3. Prick through at P_1 and Q_1 on the overlay to locate P'_1 and Q'_1 on the drawing.

Relative pole. In Fig. 5-9*a* a portion of Fig. 5-8*b* has been reproduced. AB is shown fixed in its initial position, A_1B_1. Line PQ moves relative to A_1B_1 from P_1Q_1 to $P'_2Q'_2$. Now the actual plane motion of PQ relative to AB can be replaced by a simple rotation about some point on the body containing line AB, a point called the *relative pole.*

The construction is shown in Fig. 5-9*b*. The relative pole, R_{12}, is at the intersection of the midnormals p'_{12} and q'_{12}. Midnormal p'_{12} is the midnormal of $P_1P'_2$ while q'_{12} is the midnormal of $Q_1Q'_2$. The prime on the names of these midnormals is used to emphasize the fact that they are for relative movements.

Figure 5-9*c* shows R_{12} regarded as a point on a body m, which contains line AB. Line PQ rotates about this point through an angle α_{12}, again, relative to AB.

5-5. Design of a Four-Bar Mechanism for Specified Coupler Positions

Two positions of the coupler. Two positions of a link AB are shown in Fig. 5-10. It is desired to design a mechanism which will carry the link through these two positions.

By constructing the midnormals, a_{12} and b_{12}, to A_1A_2 and B_1B_2, the pole, P_{12}, is located. The simplest mechanism would be a single link pinned to the frame at P_{12} and containing line AB. However, a four-bar mechanism will be used, where AB is the coupler.

Points A and B will be regarded as the coupler joints. Then, since point A is also the end of a crank or lever, it moves in a circular path. A_1A_2 is one chord of this path, and the center of any circle containing points A_1 and A_2 lies on the midnormal to A_1A_2, that is, a_{12}. The same reasoning applies to point B.

In the absence of further design specifications, point O_a, the center of rotation of the crank to A, can be located anywhere on midnormal a_{12}. Similarly, point O_b, the center of rotation of the crank to B, can be located anywhere on b_{12}. The completed mechanism is O_aABO_b, which has been shown in its two prescribed positions in Fig. 5-10. During the movement from the first to the second position, link O_aA moves through an angle ϕ_{12}, while line O_bB moves through ψ_{12}.

Since the positions of points O_a and O_b were chosen arbitrarily, there are an indefinite number of solutions. And because two arbitrary assumptions

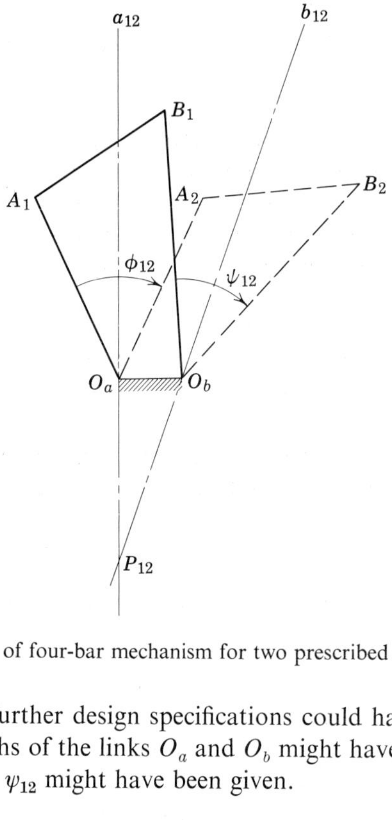

Fig. 5-10. Design of four-bar mechanism for two prescribed coupler positions.

had to be made, further design specifications could have been met. For example, the lengths of the links O_a and O_b might have been specified, or the angles ϕ_{12} and ψ_{12} might have been given.

Three positions of the coupler. The required three positions of a coupler are shown in Fig. 5-11*a*. A four-bar mechanism is to be designed, with AB as its coupler.

The reasoning follows the same pattern as for two coupler positions. Points A and B each must move in a circular path. Therefore point O_a is the center of a circle through points A_1, A_2, and A_3 while O_b is the center of a circle through B_1, B_2, B_3. And since three noncollinear points determine a circle, there is but one solution. Midnormals a_{12}, a_{23}, b_{12}, and b_{23} are constructed. O_a is at the intersection of a_{12} and a_{23}. O_b is at the intersection of b_{12} and b_{23}. They have been constructed in Fig. 5-11*b*. As a check, a_{13}

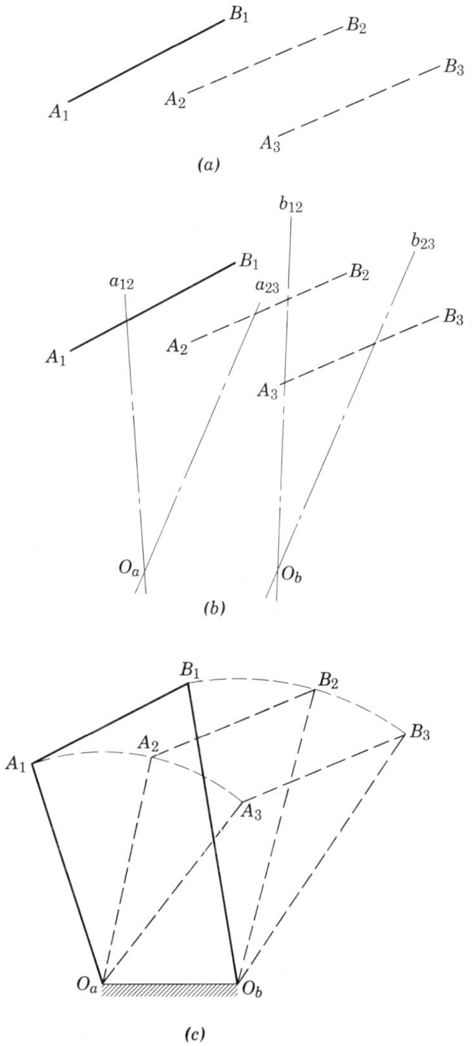

Fig. 5-11. Design of a four-bar mechanism for three coupler positions.

might be drawn. It should also pass through O_a; similarly, b_{13} should pass through O_b.

The completed mechanism in its three positions is shown in Fig. 5-11*c*.

5-6. Design of a Four-Bar Mechanism for Coordination of Angular Positions

A common requirement set a mechanism is that of coordinating the angular positions of a driven link with those of some driving link. For example, in Fig. 5-12 the angular positions of links 2 and 3 are to be coordinated, specified values of ψ corresponding to values of ϕ. The graphical techniques are based upon the notions of relative position. They will be demonstrated below by examples.

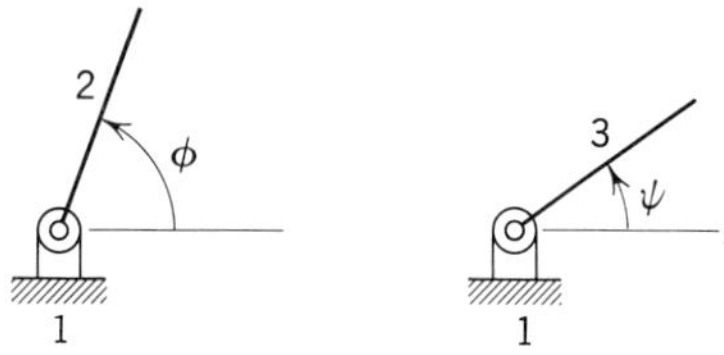

Fig. 5-12. Coordination of angular positions.

Example 5-3. Design a mechanism such that a 60° displacement of the driver will cause a 19° displacement of the driven link. Use a link mechanism.

Solution. The general problem layout is shown in Fig. 5-13*a*. Link O_aA is the driver and O_bB the driven link. The driver moves through an angle of $\Delta\phi_{12} = 60°$ from O_aA_1 while the driven link moves through an angle of $\Delta\psi_{12} = 19°$ from O_bB_1 to O_bB_2.

The number of given conditions is too small to provide a unique solution to this problem; thus some initial assumptions will have to be made. The following assumptions were made in the solution:

$$O_aO_b = 3.00$$

$$O_aA = 0.93$$

$$\phi_1 = 60°$$

The solution is shown in Fig. 5-13*b*. The reasoning employed is as follows: Since *AB* is a rigid link, *A* moves in a circular path *relative* to *B*. Therefore *B* lies at the center of a circle passing through the positions of *A* *relative* to *B*. The following steps are used in the construction illustrated in Fig. 5-13*b*:

1. Points O_a and O_b are located the assumed distance (3.00) apart.
2. The driving link, O_aA, is laid out at its initial position; that is, such that $\angle O_bO_aA_1 = \phi_1 = 60°$.
3. Link O_bB is assumed fixed in its first position, O_bB_1, although it is not known where this is. Then A_1 is the initial position of *A* relative to O_bB. It remains to find A'_2, the second position of *A* relative to O_bB.

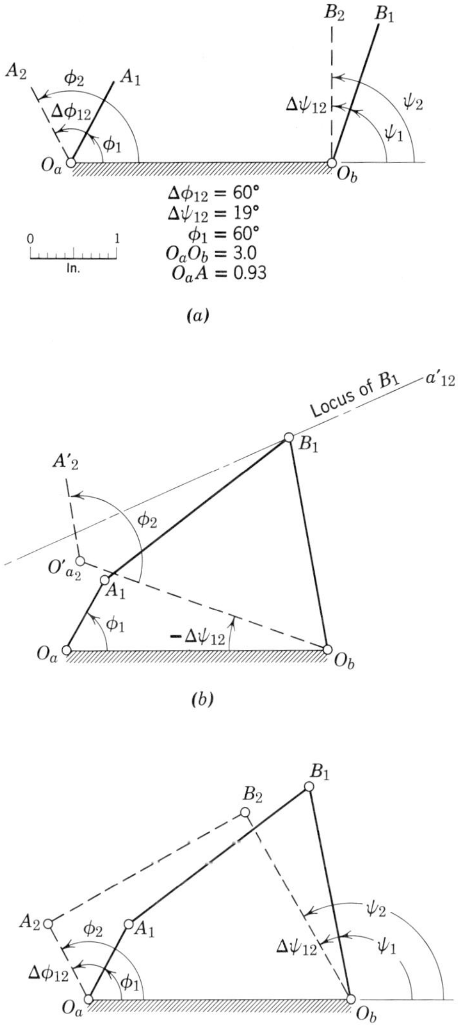

Fig. 5-13. Design of a four-bar mechanism to coordinate two angular displacements.

Since the driven link, O_bB, moves through an angle of 19° counterclockwise relative to the frame, the frame moves 19° clockwise relative to it. Thus O'_{a_2} is found by rotating O_aO_b about O_b 19° clockwise. And A'_2 is located by making $\angle A'_2O'_{a_2}O_b = \phi_2 = \phi_1 + \Delta\phi_{12} = 120°$.

4. Next the midnormal, a'_{12}, to A'_2A_1 is drawn. On this line will be found the center of any circle passing through the two relative positions of A, that is, A_1 and A'_2.

5. Point B_1 is then selected arbitrarily along a'_{12}. It is known to be specifically B_1 because link O_bB was assumed to be fixed in its first position.

6. The completed mechanism in its two positions is shown in Fig. 5-13*c*.

In the above solution three arbitrary assumptions had to be made: the length of the frame, the length of the driver, and the initial position of the driver. Therefore additional design specifications could have been met.

Example 5-4. Design a four-bar mechanism to coordinate three prescribed positions of one crank with three positions of another. A 22° movement of the driver is to be accompanied by a 21° movement of the

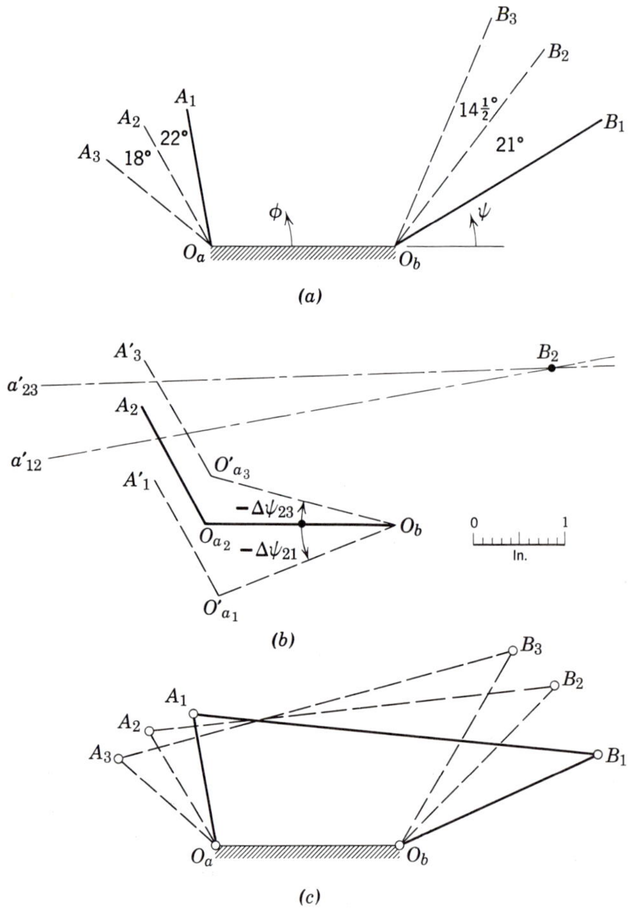

Fig. 5-14. Design of a four-bar mechanism to coordinate three positions.

driven link. Then an 18° movement of the driver is to be accompanied by a $14\frac{1}{2}°$ movement of the driven link (Fig. 5-14*a*).

Solution. Again arbitrary assumptions have to be made. They are:

$$O_aO_b = 2.00$$
$$O_aA = 1.40$$
$$\phi_1 = 100°$$

The reasoning follows the same pattern as for two positions. However, this time there are three relative positions of A, and hence but a single

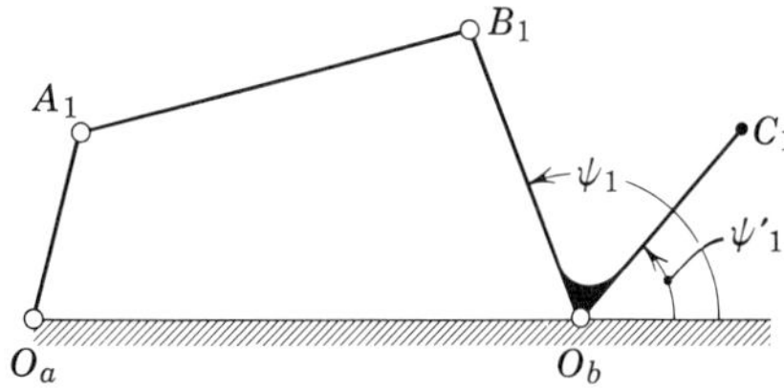

Fig. 5-15. Use of a bell-crank in a four-bar mechanism.

circle through them. There will be a unique point B_1. The construction, which is shown in Fig. 5-14*b*, involves the following steps:

1. Locate O_a and O_b the assumed distance apart.

2. Assume link O_bB fixed in one of its positions, in this case position 2. Locate A_2 such that $\angle A_2O_aO_b = \phi_2 = 100° + 22° = 122°$, and $O_aA_2 = 1.40$.

3. Next points A'_1 and A'_3 must be found. They are the positions of A relative to O_bB in first and third positions. Since the driven link would rotate 21° clockwise relative to the frame in going from second to first position, the frame is rotated 21° counterclockwise to locate O'_{a_1}. And A'_1 is located by making $\angle A'_1O'_{a_1}O_b = \phi_1 = 100°$, and $A'_1O'_{a_1} = 1.40$. Since the driven link rotates $14\frac{1}{2}°$ counterclockwise in going from its second to its third position, O'_{a_3} is found by rotating the frame $14\frac{1}{2}°$ clockwise. Then A'_3 is found by making $\angle A'_3O'_{a_3}O_b = \phi_3 = 140°$ and $A'_3O'_{a_3} = 1.40$.

4. The midnormals, a'_{12} of A'_1A_2 and a'_{23} of $A_2A'_3$, are drawn. Their intersection is the required point B, specifically point B_2 since the driven link was assumed fixed in second position.

5. The completed mechanism in its three prescribed positions is shown in Fig. 5-14*c*.

The use of a bell-crank to correct for an unfavorable orientation (from the viewpoint of the application) of the driven link is shown in Fig. 5-15.

5-7. Coupler-Point Curves

Another of the basic mechanism design problems for which the four-bar chain can provide solutions is that of constraining a point to move in a prescribed path. A common solution is to find a point on the coupler of a four-bar mechanism, usually a crank and rocker mechanism, which describes a path closely approximating the desired one.

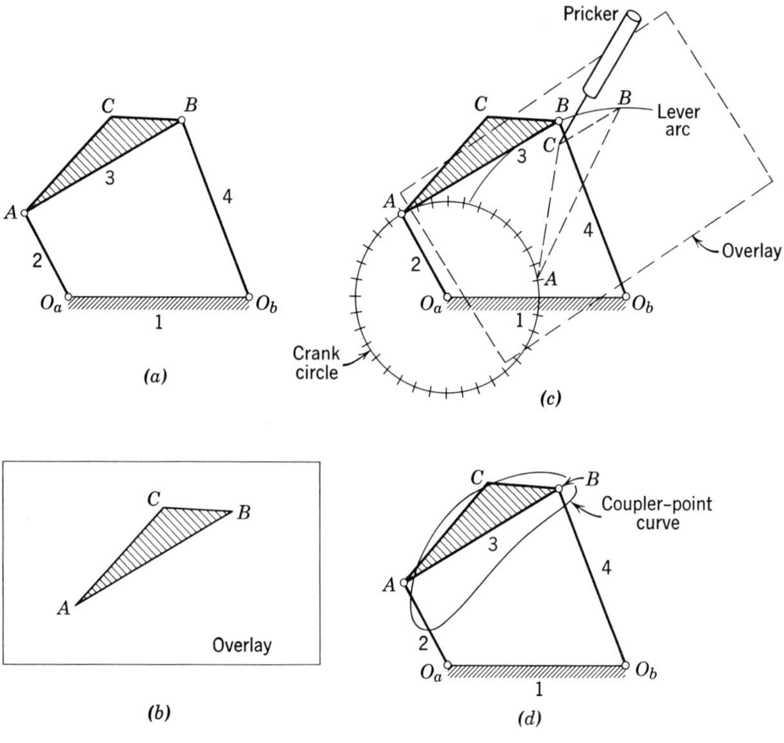

Fig. 5-16. Construction of coupler-point curve.

Constructing coupler curves. To study the possibilities of a four-bar mechanism to generate desired point paths, it is necessary to be able to plot the path of any point on the coupler, a *coupler-point curve*.

A crank and rocker mechanism is shown in Fig. 5-16*a*. It is required to plot the coupler-point curve for point *C*, a point rigidly attached to the coupler. The following procedure is convenient for this purpose:

1. Prepare a vellum overlay by tracing the coupler, *ABC*. (Fig. 5-16*b*.)
2. Draw the crank circle. This is the locus of point *A*. It is, of course,

the circle of radius O_aA drawn from O_a as center. Divide this circle into some convenient number of equal angular divisions. In Fig. 5-16*c*, 10° divisions have been used.

3. Draw the lever arc. This is the locus of point B. Draw an arc of indefinite extent and of radius O_bB from O_b as center.

4. Place the overlay over the original drawing (Fig. 5-16*c*) so that point A on the overlay coincides with one of the points marked on the crank circle. Swing the overlay to make point B fall on the lever arc. Using a pricker, mark through C on the overlay to locate a point on the original drawing.

5. Repeat the procedure for each point on the crank circle and connect the coupler curve points with an irregular curve. (Fig. 5-16*d*.)

Coupler-point curve atlas. In 1951 John A. Hrones and George L. Nelson published [Ref. 7] a set of 730 accurate drawings of the coupler-point curves for various points on the couplers of a number of crank and rocker mechanisms. The coupler-point curves are plotted as broken lines, each dash corresponding to 5° of crank rotation. By paging through this atlas, the designer can locate the particular coupler curve which will satisfy his requirements and can read off from the drawing the appropriate mechanism proportions and coupler-point location. The introductory pages of the atlas contain suggestions on its use.

Symmetrical coupler curves. If the coupler and lever of a crank and rocker mechanism are of equal length, it is possible to find a family of points on the coupler, each of which will describe a symmetrical coupler-point curve. These points fall on a circle whose radius is the length of the coupler and whose center is the coupler-lever joint. The line of symmetry of the coupler-point curve for any point on this circle will be the line through the coupler point at the location corresponding to a 180° crank position and through the lever-frame joint.

An example is shown in Fig. 5-17. A crank and rocker mechanism has been drawn at the 180° phase. A circle of radius AB has been drawn with B as center. This is the locus of points on the coupler having symmetrical coupler-point curves. Point C has been selected at random on this circle and its path plotted. The path is symmetrical about a line through point C (at 180° phase) and point O_b.

Special applications of the coupler-point curves. The number of uses to which coupler-point curves can be put seems limited only by the imagination and ingenuity of the designer. Following is one example of their use.

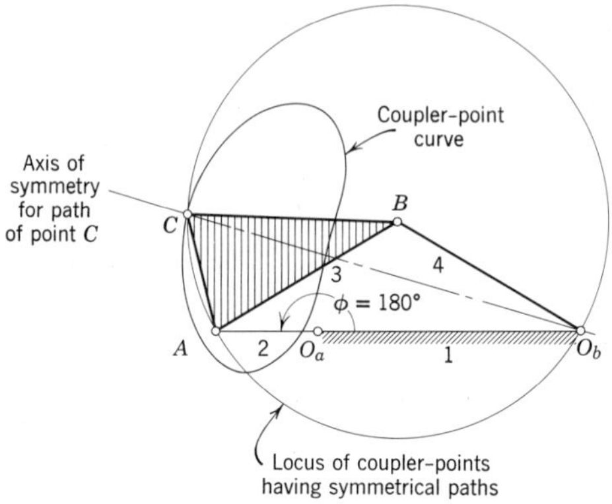

Fig. 5-17. Symmetrical coupler-point curve.

Example 5-5. Design a mechanism having as its driven member a link which will oscillate through an angle of 20° and dwell, while the mechanism crank turns through an angle of about 50°, at one of its extreme positions.

Solution. Following is a possible solution to the problem:

1. Page through the Hrones and Nelson atlas and locate a coupler-point curve having a straight segment approximately 10 dashes long (each dash corresponds to 5° of crank rotation).

2. Draw the mechanism and coupler-point curve (Fig. 5-18). In the drawing the straight segment of the coupler curve has been labeled '*ca*'.

3. Draw a line of indefinite length collinear with the straight segment of the coupler-point curve.

4. Draw a second line tangent to the coupler-point curve and making an angle of 20° with the line drawn in Step 3.

5. The intersection of the two lines just drawn is the fixed center of a slotted link. This link slides on coupler-point C and pivots about point O_d.

6. The motion timetable for the mechanism can now be determined. When the coupler-point C is at a, the crank is at a', etc. To find a', b', and c', the overlay is used. Placing the overlay such that C on the overlay coincides with a and B is on the lever arc, a pricker is used to locate a' on the original drawing coincident with A on the overlay. The motion timetable is:

Driver motion	*Driven motion*
27° to 132° 20′	Upward swing
132° 20′ to 337° 20′	Downward swing
337° 20′ to 27°	Dwell

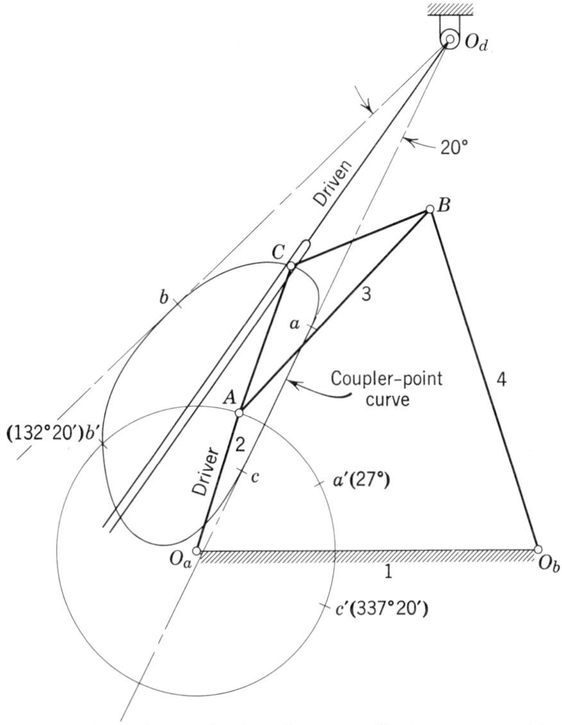

Fig. 5-18. Design of a mechanism for an oscillating output with dwell.

5-8. Sliding-Block Mechanisms

The sliding-block chain. The sliding-block chain is illustrated in Fig. 5-19. It is a four-link plane motion chain, with four lower pairs, three turning and one sliding. In Fig. 5-19*a*, an in-line chain is shown. The path of point B passes through point O_a. In Fig. 5-19*b*, the path of B is offset a distance e below O_a. This is an offset chain.

Logically the sliding-block chain may be regarded as an evolution of a Class 1 four-bar chain. This is demonstrated in Fig. 5-20. A Class 1 four-bar chain is illustrated in Fig. 5-20*a*. Relative to link 1, link 4 has oscillatory motion. Link 4 can be replaced by a slider moving in a circular slot in link 1. This is shown in Fig. 5-20*b*. The four-bar chains of Figs. 5-20*a* and 5-20*b* are kinematically equivalent. If the radius of the slot is allowed to increase without limit, the slot becomes straight, and the four-bar chain has been evolved into a sliding block chain, Fig. 5-20*c*.

Inversions. The four-possible inversions of the sliding-block chain are shown in Fig. 5-21. Inversion 1 is the familiar slider-crank mechanism, in

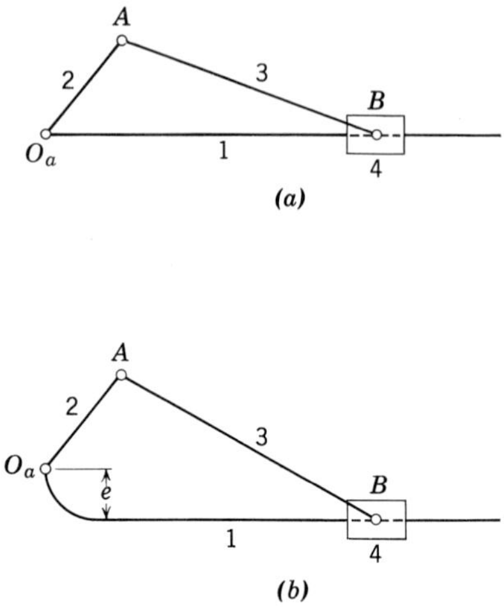

Fig. 5-19. The sliding-block chain. (*a*) In-line. (*b*) Offset.

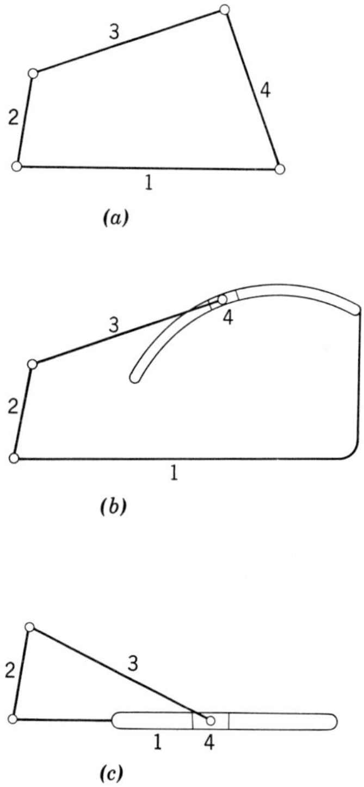

Fig. 5-20. Evolution of the sliding-block chain from the four-bar chain.

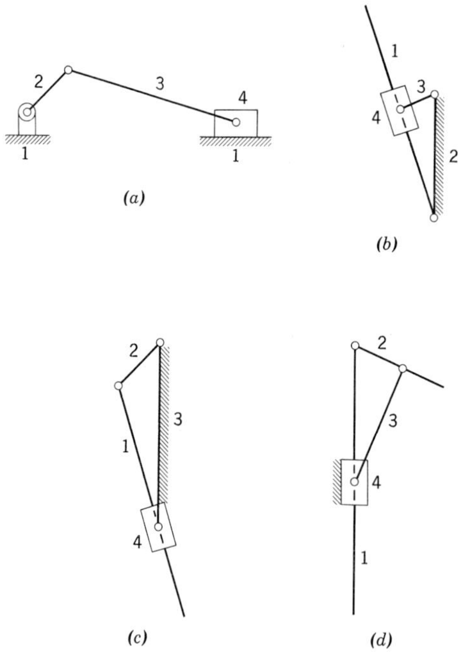

Fig. 5-21. Inversions of the sliding-block chain. (*a*) First inversion. (*b*) Second inversion (*c*) Third inversion. (*d*) Fourth inversion.

which the guide, link 1, is fixed. This mechanism is used in engines and compressors.

In inversion 2 the crank, link 2, is fixed. A common application of this mechanism is as a quick-return for machines such as the shaper.

Inversion 3 has found use in oscillating marine steam engines. Here link 3 is fixed.

Perhaps the least used of the four inversions is inversion 4, in which the block is the fixed link. If link 2 is extended and used as a handle, this mechanism can be used as a simple pump.

The sections which follow will be devoted to a discussion of the motion characteristics of the first inversion, the slider-crank mechanism.

5-9. Graphical Determination of Slider Displacement

The construction is illustrated in Fig. 5-22. Here point B, the center of the wristpin, is used as a reference point for the slider. It is required to find the position of B for a specified phase, θ, of the crank. The following simple construction is used:

1. Draw the crank at the given phase, and draw the path of travel of point B.
2. With A as center, strike an arc of radius AB intersecting the line of travel of B at point B. AB is the connecting rod length.

When the wristpin, B, is farthest from the crank center, O, it (and the slider) is said to be at *top dead center position*. When nearest O, it is at *bottom dead center*. It is conventional among engine designers to define the slider displacement as its distance, s, from top dead center.

For either the in-line or the offset slider-crank mechanism, top dead center position occurs when the crank and connecting rod are stretched out, and bottom dead center when they overlap. For the in-line mechanism (Fig. 5-22*a*), top dead center position occurs when

$$OB = L + R$$

and bottom dead center when

$$OB = L - R$$

Thus the total stroke of the slider is

$$\text{stroke} = (L + R) - (L - R) = 2R$$

For the offset mechanism, the situation is more complicated, as Fig. 5-22*b* indicates. The total stroke for this case can be found quickly by locating top and bottom dead center positions graphically. Top dead center is at the intersection of an arc of radius $L + R$, struck from O, with the line of travel of B. And bottom dead center is at the intersection of an arc of radius $L - R$, struck from O, with the line of travel of B. Once these two points are located, the stroke can be measured directly on the drawing.

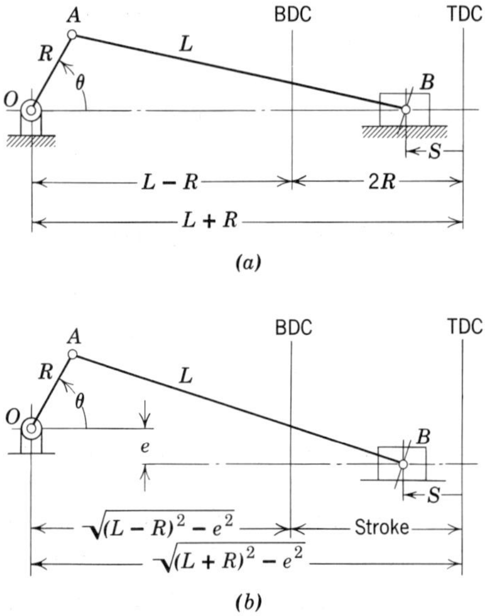

Fig. 5-22. Graphical determination of slider displacement.

5-10. Graphical Determination of Velocities

Slider velocity. A rapid method for determining the slider velocity at any phase is shown in Fig. 5-23*a*. It is assumed that the mechanism has already been laid out for the phase in question. The method is as follows:

1. Draw a line from O in a direction perpendicular to the direction of travel of B.
2. Label the point where this line crosses the connecting rod, or connecting rod extended, 'C'.
3. Then, to the same scale, OA represents V_A, OC represents V_B, and CA represents $V_{A/B}$.

Proof

1. The velocity polygon has been drawn in Fig. 5-23*b* by use of the methods of Chapter 3.

2. Since

$$Oa \perp OA$$
$$ab \perp AC$$
$$Ob \perp OC$$

it follows that triangles OCA and Oba are similar.

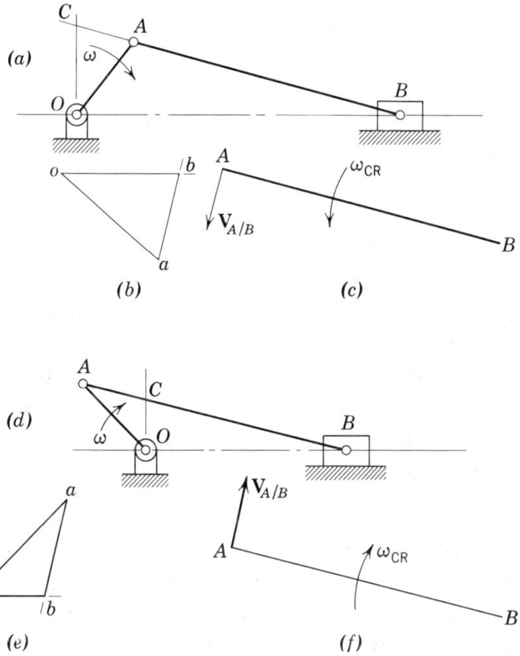

Fig. 5-23. Graphical determination of velocities in a slider-crank mechanism.

3. Since oa, ob, and ba represent V_A, V_B, and $V_{A/B}$ respectively to some common scale, then also OA, OC, and CA represent V_A, V_B, and $V_{A/B}$ respectively to some other common scale.

It remains to determine this second scale. Drawing distance OA represents the crank length, R, to the distance scale, k_s, and the crankpin speed, V_A, to some speed scale, k_v. That is,

$$k_s = \frac{R}{OA}$$

and

$$k_v = \frac{V_A}{OA}$$

Now

$$V_A = R\omega$$

and from the first scale equation,

$$OA = \frac{R}{k_s}$$

Therefore,
$$k_v = \omega \cdot k_s \tag{5-7}$$

Now that the speed scale has been found, the findings thus far can be summarized in a set of equations:

$$V_A = OA \cdot \omega \cdot k_s \tag{5-8}$$

$$V_B = OC \cdot \omega \cdot k_s \tag{5-9}$$

$$V_{A/B} = CA \cdot \omega \cdot k_s \tag{5-10}$$

where R = crank length, ft
ω = crank speed, rad/sec
V_A = crankpin speed, ft/sec
V_B = slider speed, ft/sec
$V_{A/B}$ = speed of A relative to B, ft/sec
OA = drawing distance representing crank length to scale k_s and crankpin speed to scale k_v, in.
OC = drawing distance representing slider speed to scale k_v, in.
CA = drawing distance representing speed of A relative to B to scale k_v, in.
k_s = distance scale, ft/in.
$k_v = \omega k_s$ = speed scale, ft/sec/in.

Although special rules could be applied which would determine in a mechanical fashion the direction of the slider's velocity, it is simpler to determine the direction by inspection of the drawing. For example, in Fig. 5-23*a*, with the crank moving clockwise, it is clear that, for the phase shown, the slider is moving to the right.

Connecting-rod velocity. The angular speed of a rigid body in plane motion can be found by dividing the relative speed of two points on the body by the distance between them (Art. 3-11). The construction described above, with Eq. 5-10, furnishes the relative speed of points A and B, which are rigidly attached to the connecting rod. Thus,

$$\omega_{\text{CR}} = \frac{V_{A/B}}{L}$$

But
$$V_{A/B} = CA \cdot \omega \cdot k_s \tag{5-10}$$

and
$$L = AB \cdot k_s$$

Therefore,
$$\omega_{\text{CR}} = \frac{V_{A/B}}{L} = \frac{CA \cdot \omega \cdot k_s}{AB \cdot k_s}$$

or
$$\omega_{CR} = \frac{CA}{AB}\,\omega \tag{5-11}$$

where ω_{CR} = connecting rod speed, rad/sec
ω = crank speed, rad/sec

A simple means for determining the angular speed of the connecting rod is now available. It remains to state a rule for determining its direction. The following rule is convenient: *If point C is between A and B, the crank and connecting rod are turning in the same direction; if point C is not, they turn in opposite directions.*

The rule is illustrated in Fig. 5-23. Point C lies outside the segment AB in Fig. 5-23*a*, implying that the connecting rod and crank turn in opposite directions. And, since the crank is turning in a clockwise direction, this implies that the connecting rod is turning in a counterclockwise direction.

To check the validity of the rule, the velocity polygon has been drawn in Fig. 5-23*b*. An inspection of the polygon will show that $\mathbf{V}_{A/B}$ is down and to the left. This implies, as Fig. 5-23*c* demonstrates, that the connecting rod is turning in a counterclockwise direction. The rule checks.

A further application of the rule is made in Fig. 5-23*d*, where the same slider-crank mechanism is shown at a different phase. Here point C falls between A and B, implying that the crank and connecting rod turn in the same direction. Because the crank turns clockwise, the connecting rod must also.

To check the rule once more, the velocity polygon for this phase has been constructed in Fig. 5-23*e*. From an inspection of the polygon it is evident that $\mathbf{V}_{A/B}$ is up and to the right. Thus, as Fig. 5-23*f* demonstrates, the connecting rod is turning in a clockwise direction.

To further check the rule to his own satisfaction, the reader should sketch velocity polygons for each phase, assuming that the crank is turning in a counterclockwise direction. Using the velocity polygons to determine the direction of rotation of the connecting rod for each phase, he can then see if the rule gives the correct result.

Mechanical advantage. The approach to the analysis of mechanical advantage in the slider-crank mechanism will follow the same pattern as for the four-bar mechanisms [Art. 5-3].

If friction is ignored, the input and output power are the same. That is,

$$P_i = P_o$$

or
$$F_i V_i = F_o V_o$$

The mechanical advantage (MA) is

$$\text{MA} = \frac{F_o}{F_i} = \frac{V_i}{V_o}$$

Using the crankpin as the input point and the slider as the output, the crankpin and slider speeds are

$$V_A = OA \cdot \omega \cdot k_s \tag{5-8}$$

$$V_B = OC \cdot \omega \cdot k_s \tag{5-9}$$

Therefore, the mechanical advantage is

$$\text{MA} = \frac{V_A}{V_B} = \frac{OA}{OC} \tag{5-12}$$

If the slider is the input member, then the mechanical advantage is

$$\text{MA} = \frac{OC}{OA} \tag{5-13}$$

Example 5-6. An in-line slider-crank mechanism (Fig. 5-24) used in a reciprocating compressor has a crank-length of 4 in. and a connecting rod

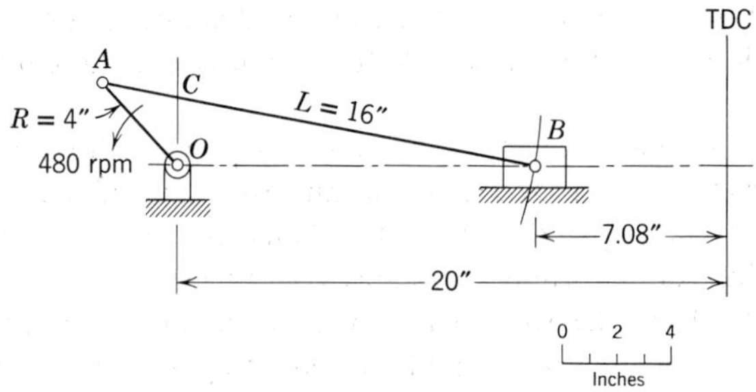

Fig. 5-24. Solution of Example 5-6.

length of 16 in. The crank turns counterclockwise at 480 rpm. Find the slider velocity and displacement, the connecting rod angular velocity, and the mechanical advantage at the 135° phase.

Solution

1. The mechanism was originally drawn at the distance scale $k_s = \frac{1}{3}$ ft/in. (quarter size).

2. Top dead center position has been located at a distance of $L + R =$ 20 in. from O.

3. By measurement the displacement is 7.08 in.

4. Drawing distance OC was 0.58 in. And

$$\omega = \frac{2\pi N}{60}$$

$$= \frac{2\pi(480)}{60}$$

$$= 50.27 \text{ rad/sec}$$

Using Eq. 5-9,

$$V_B = OC \cdot \omega \cdot k_s$$

$$V_B = (0.58)(50.27)(1/3)$$

$$= 9.72 \text{ ft/sec}$$

5. To find the angular speed of the connecting rod, Eq. 5-11 is used.

$$\omega_{\text{CR}} = \frac{CA}{AB}\,\omega \qquad (5\text{-}11)$$

The drawing distances were $CA = 0.73$ in. and $AB = 4.0$ in. Therefore,

$$\omega_{\text{CR}} = \frac{0.73}{4.00}(50.27)$$

$$= 9.17 \text{ rad/sec}$$

Since point C falls between A and B, the connecting rod is turning in the same direction as the crank, counterclockwise.

6. At this phase the mechanical advantage is

$$\text{MA} = \frac{OA}{OC} \qquad (5\text{-}12)$$

$$= \frac{1.00}{0.58}$$

$$= 1.72$$

5-11. Graphical Determination of Accelerations

The Ritterhaus construction. This construction is a short-cut method for determining the accelerations in a slider-crank mechanism. And it, like the velocity construction of the previous article, involves constructions on the mechanism layout itself, making unnecessary the construction of a separate acceleration polygon. The method is illustrated in Figs. 5-25*a* and 5-25*c* for two phases of a slider-crank mechanism. The following steps are necessary:

1. From O draw a line perpendicular to OB.
2. Locate the point C where the connecting rod, extended if necessary, crosses the line of Step 1.

3. From C draw a line parallel to OB.

4. Locate the point D where the crank, extended if necessary, intersects the line of Step 3.

5. From D draw a line perpendicular to OB and intersecting the connecting rod at E.

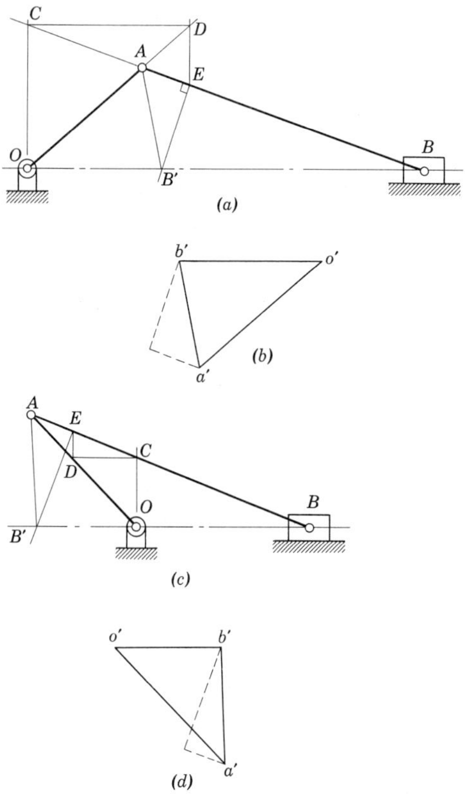

Fig. 5-25. The Ritterhaus construction.

6. From E draw a line perpendicular to the connecting rod to intersect OB at B'.

The reader should follow through the sequence of steps both in Fig. 5-25*a* and in Fig. 5-25*c*. The method is the same for each case, though the final results are very different in appearance.

Interpretation. For each of the two phases of the slider-crank mechanism in Fig. 5-25, the acceleration polygon has also been drawn for purposes of comparison. It can be proved that the figure formed by the lines joining points O, A, E, and B' in the Ritterhaus construction is similar to the corresponding acceleration polygon. However, the figure is turned 180° out of phase with the acceleration polygon.

Drawing distance OA represents the crank length, R, to the scale k_s; but, as a segment of the Ritterhaus construction, it also represents the normal acceleration, $A_A{}^n$, of the crankpin to some yet undetermined scale, k_a, the acceleration scale. That is,

$$k_s = \frac{R}{OA}$$

and

$$k_a = \frac{A_A{}^n}{OA}$$

But

$$A_A{}^n = R\omega^2$$

and, from the first scale equation above,

$$OA = \frac{R}{k_s}$$

Therefore,

$$k_a = \frac{R\omega^2}{R/k_s}$$

or

$$k_a = \omega^2 k_s \tag{5-14}$$

Equation 5-14 provides a key to the interpretation of the Ritterhaus construction, for it determines to what scale the segments of the Ritterhaus construction represent the various accelerations.

TABLE 5-2

Vector	Represented by Drawing Distance	Sense
$\mathbf{A}_A{}^n$	AO	From A toward O
$\mathbf{A}_B$	$B'O$	From B' toward O
$\mathbf{A}^n_{A/B}$	AE	From A toward E
$\mathbf{A}^t_{A/B}$	EB'	From E toward B'
$\mathbf{A}_{A/B}$	AB'	From A toward B'

A comparison of the figure $OAEB'$ in the Ritterhaus construction with the corresponding acceleration polygon leads to the rules of correspondence shown in Table 5-2.

The magnitude of the angular acceleration of a body having plane motion can be found by dividing the magnitude of the tangential component of the relative acceleration of any two points on the body by the distance between them (Art. 4-2). For the connecting rod then,

$$\alpha_{CR} = \frac{A^t_{A/B}}{L}$$

But $A^t_{A/B}$ is represented to the scale k_a by drawing distance EB', and L is represented to the scale k_s by drawing distance AB. That is,

$$A^t_{A/B} = EB' \cdot k_a$$

or, using Eq. 5-14,

$$A^t_{A/B} = EB' \cdot \omega^2 \cdot k_s$$

and

$$L = AB \cdot k_s$$

Therefore

$$\alpha_{CR} = \frac{EB' \cdot \omega^2 \cdot k_s}{AB \cdot k_s}$$

or

$$\alpha_{CR} = \frac{EB'}{AB} \cdot \omega^2 \tag{5-15}$$

When E is above B', the angular acceleration of the connecting rod is counterclockwise; when E is below B', the direction is clockwise. This rule assumes that the mechanism is being viewed as in Fig. 5-25. If viewed from the other side, that is in such a way that O is to the right of B, the directions would appear reversed.

Slider acceleration at 0° and 180° phases. The Ritterhaus construction is not applicable when the slider-crank mechanism is at its top dead center or bottom dead center positions. For these positions the general methods of acceleration analysis of Chapter 4, particularly the graphical method for determining normal acceleration of Art. 4-4, must be used.

A slider-crank mechanism in its two extreme positions is shown in Fig. 5-26. For each case,

$$\mathbf{A}_B = \mathbf{A}_A + \mathbf{A}_{B/A}$$

or

$$\mathbf{A}_B = \mathbf{A}_A{}^n + \mathbf{A}_A{}^t + \mathbf{A}^n_{B/A} + \mathbf{A}^t_{B/A}$$

Since ω is assumed to be constant, $\mathbf{A}_A{}^t = 0$. And, since $\mathbf{A}_B$ and $\mathbf{A}^n_{B/A}$ at these extreme positions must be collinear, $\mathbf{A}^t_{B/A} = 0$. Hence,

$$\mathbf{A}_B = \mathbf{A}_A{}^n + \mathbf{A}^n_{B/A}$$

Further, since

$$\mathbf{V}_B = \mathbf{V}_A + \mathbf{V}_{B/A}$$

and, at these extreme positions, $\mathbf{V}_B = 0$, it follows that

$$\mathbf{V}_{B/A} = -\mathbf{V}_A$$

or

$$\mathbf{V}_{A/B} = \mathbf{V}_A$$

These considerations justify the following combined construction, which is simply an application of the graphical method for determining normal acceleration explained in Art. 4-4. The construction is shown in Fig. 5-26*a*

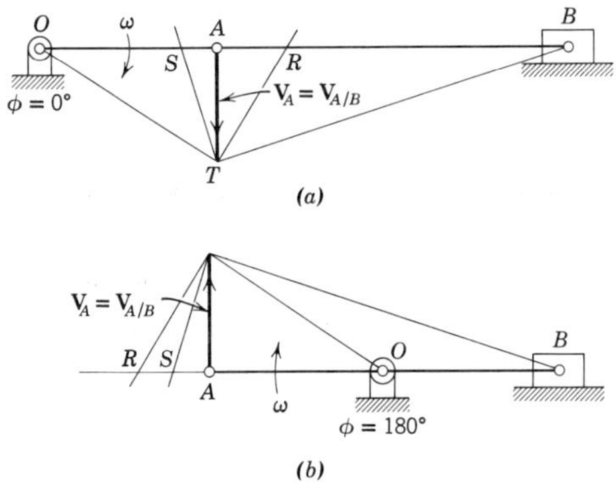

Fig. 5-26. Slider acceleration at 0° and 180° phases.

for the 0° phase and in Fig. 5-26*b* for the 180° phase. For each phase the method is the same.

1. At A lay out to an arbitrary velocity scale, k_v, the crankpin velocity, $\mathbf{V}_A$. Call its terminus 'T'.
2. From T draw a line perpendicular to OT and intersecting OB, extended if necessary, at R.
3. From T draw a line perpendicular to BT and intersecting OB, extended if necessary, at S.
4. Then drawing distance RS represents A_B to the scale

$$k_a = \frac{k_v^{\,2}}{k_s} \tag{4-10}$$

and the sense of $\mathbf{A}_B$ is from R toward S.

Proof

1. By the method and proof of Art. 4-4, drawing distance RA represents, to the scale k_a, $A_A^{\,n}$. And the sense of $\mathbf{A}_A^{\,n}$ is that of $\overrightarrow{RA}$ (necessarily toward the center of rotation of the crank, point O). Therefore,

$$\mathbf{A}_A^{\,n} = \overrightarrow{RA} \cdot k_a$$

2. Similarly, by the method of Art. 4-4, drawing distance SA represents, to the scale k_a, $A^n_{A/B}$ or $A^n_{B/A}$. The sense of $\mathbf{A}^n_{B/A}$ is from B toward A and therefore the same as the sense of $\overrightarrow{AS}$. That is,

$$\mathbf{A}^n_{B/A} = \overrightarrow{AS} \cdot k_a$$

3. And finally,

$$\mathbf{A}_B = \mathbf{A}_A{}^n + \mathbf{A}^n_{B/A}$$

$$= \overrightarrow{RA} \cdot k_a + \overrightarrow{AS} \cdot k_a$$

or

$$\mathbf{A}_B = \overrightarrow{RS} \cdot k_a$$

That is, A_B is represented to the scale k_a by drawing distance RS, and the sense of $\mathbf{A}_B$ is the same as that of $\overrightarrow{RS}$, from R toward S.

Comparison of slider motion and simple harmonic motion. It is often useful to plot the motion characteristics of a mechanism in order to study

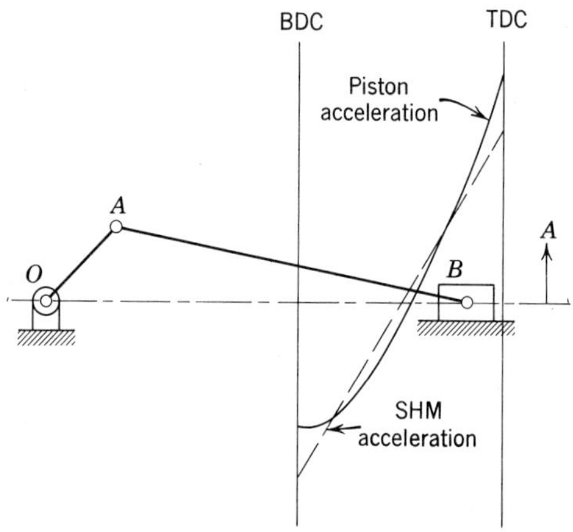

Fig. 5-27. Comparison of slider acceleration and simple harmonic motion acceleration.

their patterns. In Fig. 5-27 an acceleration-space curve has been plotted for the slider of a slider-crank mechanism. This curve is a plot of the acceleration of the slider as ordinate at the corresponding slider position.

In addition, an acceleration-space curve has been plotted for a Scotch yoke having the same crank radius and crank speed. The acceleration-space curve for the Scotch yoke is a straight line. In a Scotch yoke the yoke moves with simple harmonic motion. And in simple harmonic motion the acceleration is always proportional to displacement (Art. 2-14).

The acceleration curve for the slider of the slider-crank mechanism deviates sharply from that for simple harmonic motion. This deviation is the result of what is often called the *connecting rod effect*. As the connecting rod is lengthened, for any given crank radius, the slider acceleration curve more nearly approaches that for simple harmonic motion.

Example 5-7. An in-line slider-crank mechanism used in a reciprocating compressor (the mechanism of Example 5-6) has a crank length of 4 in. and a connecting rod length of 16 in. The crank turns counterclockwise at

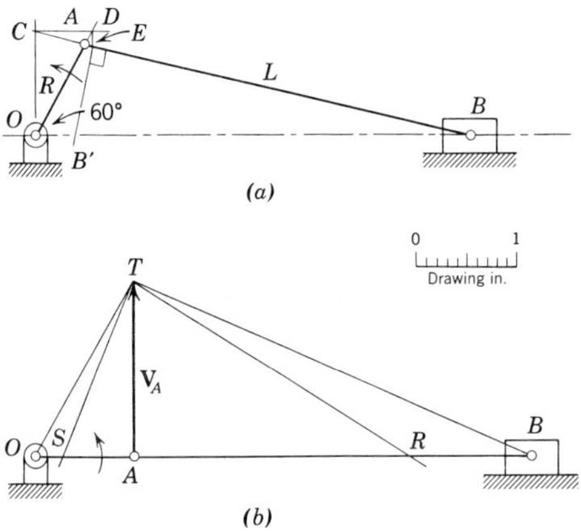

Fig. 5-28. Solutions for Example 5-7.

480 rpm. Find the slider acceleration and the angular acceleration of the connecting rod at the 60° and 0° phases.

Solution

a. *60° phase*

1. The mechanism has been drawn quarter size ($k_s = \frac{1}{3}$ ft/in.) at its 60° phase in Fig. 5-28*a*.

2. The Ritterhaus construction is carried out in Fig. 5-28*a* also. The required drawing distances are $B'O = 0.375$ in., $EB' = 0.88$ in., and $AB = 4.00$ in.

3. The angular speed of the crank is

$$\omega = \frac{2\pi N}{60}$$

$$= \frac{2\pi(480)}{60}$$

$$= 50.27 \text{ rad/sec}$$

The acceleration scale is

$$k_a = \omega^2 \cdot k_s$$
$$= (50.27)^2 \cdot (1/3)$$
$$= 842.3 \frac{\text{ft/sec}^2}{\text{in.}}$$

4. Therefore,

$$A_B = B'O \cdot k_a$$
$$= (0.375)(842.3)$$
$$= 315.9 \text{ ft/sec}^2$$

and

$$\alpha_{\text{CR}} = \frac{EB'}{AB}\,\omega^2$$
$$= \frac{0.88}{4.00}(50.27)^2$$
$$= 556 \text{ rad/sec}^2$$

The sense of $\mathbf{A}_B$ is to the left, and the connecting rod is accelerating in a counterclockwise direction.

b. 0° *phase*

1. The mechanism has been drawn at quarter size ($k_s = \frac{1}{3}$ ft/in.) in Fig. 5-28*b*.

2. $\mathbf{V}_A$ has been laid out at A to the scale $k_v = 10$ ft/sec/in. The magnitude of $\mathbf{V}_A$ is

$$V_A = R\omega$$
$$= (4/12)(50.27)$$
$$= 16.76 \text{ ft/sec}$$

3. Point R is located by drawing a line from T, the terminus of $\mathbf{V}_A$, perpendicular to OT and intersecting OB at R. Point S is located by drawing a line from T, perpendicular to BT, and intersecting OB at S. Drawing distance RS is measured to be 3.5 in.

4. The acceleration scale, k_a, is

$$k_a = \frac{k_v^{\,2}}{k_s} \tag{4-10}$$
$$= \frac{(10)^2}{1/3}$$
$$= 300 \frac{\text{ft/sec}^2}{\text{in.}}$$

Therefore,

$$\begin{aligned} A_B &= RS \cdot k_a \\ &= (3.50)(300) \\ &= 1050 \text{ ft/sec}^2 \end{aligned}$$

The sense of A_B is toward the left.

5. At the 0° and 180° phases, $A^t_{A/B} = 0$ and, hence, $\alpha_{\text{CR}} = 0$.

5-12. Slider-Crank Motion Formulas

When greater accuracy is required than can be obtained by graphical methods, resort is made to equations of motion for the mechanism. Frequently these equations are very complicated and their solution tedious. However, modern methods of machine computation make the use of equations practical in investigating the motion characteristics of a mechanism in those exceptional cases where the expense is justifiable.

In this article, a list of formulas for the motion of the slider and the connecting rod in a slider-crank mechanism will be given. Since good drafting techniques can achieve accuracy comparable to that obtained in slide-rule computations, a desk calculator or digital computer should be used where the motive for using these formulas is higher accuracy.

Slider motion. A drawing of the slider-crank mechanism exhibiting the necessary notation is given in Fig. 5-29. In the formulas that follow, the crank velocity is assumed constant.

Let
$$\frac{L}{R} - n$$

Since
$$h = R \sin \theta = L \sin \phi$$

then
$$\sin \phi = \frac{\sin \theta}{n} \qquad (5\text{-}16)$$

Now
$$s = L + R - R \cos \theta - L \cos \phi$$

or
$$s = R(1 - \cos \theta) + L(1 - \cos \phi)$$

A fundamental identity of plane trigonometry is

$$\cos \phi = (1 - \sin^2 \phi)^{1/2}$$

Substituting from Eq. 5-16,

$$\cos \phi = \left(1 - \frac{\sin^2 \theta}{n^2}\right)^{1/2}$$

Therefore,
$$s = R(1 - \cos \theta) + L\left[1 - \left(1 - \frac{\sin^2 \theta}{n^2}\right)^{1/2}\right] \qquad (5\text{-}17)$$

Equation 5-17 furnishes an exact means for calculating the slider displacement. However, it is a very cumbersome equation to handle; consequently, an approximate form is used for most engineering applications:

$$s = R(1 - \cos\theta) + \frac{R^2}{2L}\sin^2\theta \qquad (5\text{-}18)$$

The slider displacement, s, is measured from top dead center position, as indicated in Fig. 5-29. The value of s is always positive.

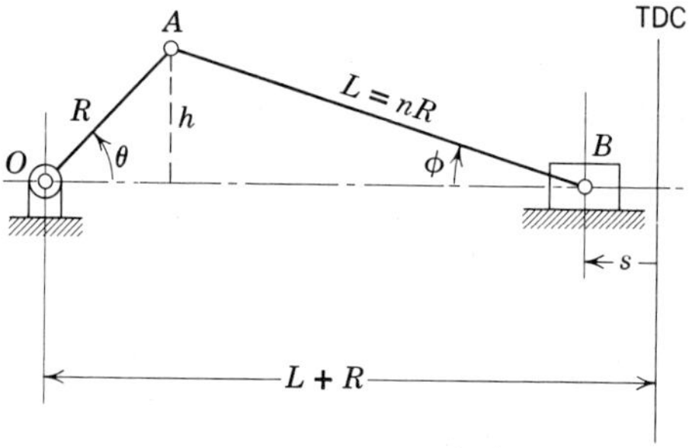

Fig. 5-29. Notation for slider-crank formulas.

Approximate equations for slider velocity and acceleration are

$$V_B = R\omega\left(\sin\theta + \frac{\sin 2\theta}{2n}\right) \qquad (5\text{-}19)$$

$$A_B = R\omega^2\left(\cos\theta + \frac{\cos 2\theta}{n}\right) \qquad (5\text{-}20)$$

In Eqs. 5-19 and 5-20, a positive value of ω corresponds to counterclockwise direction. Positive values of V_B or A_B indicate that the slider has a velocity or acceleration towards the crankpin, O.

Connecting rod motion. The following formulas are exact equations for calculating the angular velocity and acceleration of the connecting rod.

$$\omega_{\text{CR}} = -\frac{\omega\cos\theta}{(n^2 - \sin^2\theta)^{1/2}} \qquad (5\text{-}21)$$

$$\alpha_{\text{CR}} = \frac{\omega^2(n^2 - 1)\sin\theta}{(n^2 - \sin^2\theta)^{3/2}} \qquad (5\text{-}22)$$

In Eqs. 5-21 and 5-22 a positive value of an angular velocity or an angular acceleration implies a counterclockwise direction.

Crank angle for maximum slider velocity. Exact methods for determining the crank angle at which maximum slider velocity occurs are given in papers by Burl and Unwin (Refs. 3 and 11). Unwin (Ref. 11) also gives a good approximation. A very simple formula which gives the crank angle for maximum slider velocity has been devised recently by Ip and Price (Ref. 4). Their formula is applicable for values of n above 1.5.

$$\cos\theta = \frac{1}{(n^2+3)^{1/2}} \tag{5-23}$$

The Ip and Price formula is within about 1 minute of the correct angle at $n = 4.0$.

Example 5-8. An in-line slider-crank mechanism has a crank length of 4 in. and a connecting rod length of 16 in. The crank turns counterclockwise at 480 rpm. Find the slider displacement, velocity, and acceleration, as well as the angular velocity and acceleration of the connecting rod, at the 60° phase.

Solution. Equations 5-18 to 5-22 will be used. In addition to the values of R and θ, which are given, the values of ω and n are required.

$$n = \frac{L}{R} = \frac{16}{4} = 4$$

$$\begin{aligned}\omega &= \frac{2\pi N}{60}\\ &= \frac{2\pi(480)}{60}\\ &= 50.266 \text{ rad/sec}\end{aligned}$$

The required quantities can now be computed.

$$s = R(1-\cos\theta) + \frac{R^2}{2L}\sin^2\theta \tag{5-18}$$

$$\begin{aligned}&= 4(1-\cos 60) + \frac{16}{32}\sin^2 60\\ &= 2.375 \text{ in.}\end{aligned}$$

$$V_B = R\omega\left(\sin\theta + \frac{\sin 2\theta}{2n}\right) \tag{5-19}$$

$$\begin{aligned}&= (4/12)(50.266)\left(\sin 60 + \frac{\sin 120}{8}\right)\\ &= 16.324 \text{ ft/sec}\end{aligned}$$

$$A_B = R\omega^2\left(\cos\theta + \frac{\cos 2\theta}{n}\right) \tag{5-20}$$

$$\begin{aligned}&= (4/12)(50.266)^2\left(\cos 60 + \frac{\cos 120}{4}\right)\\ &= 315.83 \text{ ft/sec}^2\end{aligned}$$

This same mechanism was analyzed in Example 5-7. The slider acceleration obtained graphically was 315.9 ft/sec².

$$\omega_{\text{CR}} = -\frac{\omega \cos \theta}{(n^2 - \sin^2 \theta)^{1/2}} \tag{5-21}$$

$$= -\frac{(50.266)(\cos 60)}{(16 - \sin^2 60)^{1/2}}$$

$$= -6.4359 \text{ rad/sec}$$

$$\alpha_{\text{CR}} = \frac{\omega^2(n^2 - 1) \sin \theta}{(n^2 - \sin^2 \theta)^{3/2}} \tag{5-22}$$

$$= \frac{(50.266)^2(16 - 1) \sin 60}{(16 - \sin^2 60)^{3/2}}$$

$$= 551.15 \text{ rad/sec}^2$$

In Example 5-7 the graphical method yielded a value of 556 rad/sec².

5-13. Design of a Mechanism to Coordinate an Angular and a Linear Displacement

The approach to linkage synthesis used in Art. 5-6 can be applied here also. Figure 5-30 shows the nature of the problem. In this mechanism an

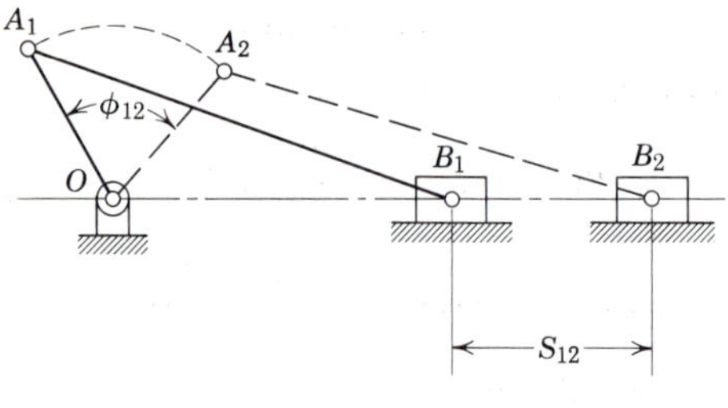

Fig. 5-30. Mechanism to coordinate an angular displacement and a linear displacement.

angular displacement of the crank, ϕ_{12}, is accompanied by a linear displacement of the slider, s_{12}. The aim of this article is to present a technique by which such a mechanism can be designed. The given quantities are ϕ_{12} and s_{12}. The technique will be demonstrated by means of an example.

Example 5-9. Design a mechanism in which a clockwise rotation of 50° by the crank is accompanied by a slider movement of 1.6 in. to the right.

Solution. The solution is shown in Fig. 5-31.

1. First points O, B_1, and B_2 must be laid out. OB_1 is arbitrarily chosen to be 3.0 in. Then B_1B_2 is made equal to s_{12}, that is, 1.6 in.

2. The crank is regarded as fixed in its first position. Since the crank moves 50° clockwise relative to the frame in going from first to second position, the frame must be moved 50° counterclockwise relative to the crank. Then B'_2 is located along the rotated frame line at a distance $OB'_2 = OB_2$.

3. B_1 and B'_2 are the relative positions of B. But relative to point A, point B moves in a circular path of which A is the center. Midnormal b'_{12} is drawn to chord $B_1B'_2$. This midnormal is the locus of all possible positions of A_1.

4. A_1 is selected arbitrarily on b'_{12}. The completed mechanism is OA_1B_1. The mechanism has been drawn in its two positions.

In the above construction two arbitrary choices were made: the length of OB_1 and the position of A_1 on the midnormal. Therefore additional design specifications could have been met. The length of the crank or of the connecting rod could have been specified, or the initial position of the crank, etc. In particular, the arbitrary quantities could be varied by the designer until he arrived at a design having a favorable range of mechanical advantage over the range of motion.

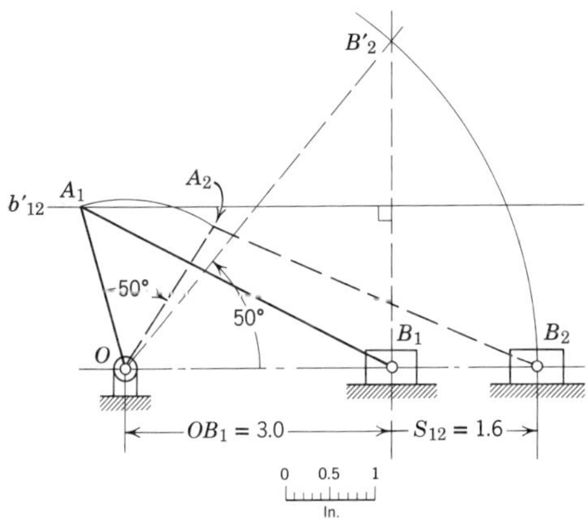

Fig. 5-31. Solution of Example 5-9.

5-14. Miscellaneous Mechanisms

For broad surveys of the numerous mechanisms which have proved practical in practice, the reader is advised to consult Jones (Ref. 8) and Horton (Ref. 6), as well as the valuable catalog by function of mechanisms given by Beggs (Ref. 1).

Straight-line mechanisms. A number of straight-line mechanisms have been based upon the four-bar chain. Among them are mechanisms devised by Watt, Evans, and Tchebisheff. Watt's mechanism is shown in Fig. 5-32. Recommendations for optimum proportions of this mechanism are given in a survey of straight-line mechanisms by Rappaport (Ref. 10).

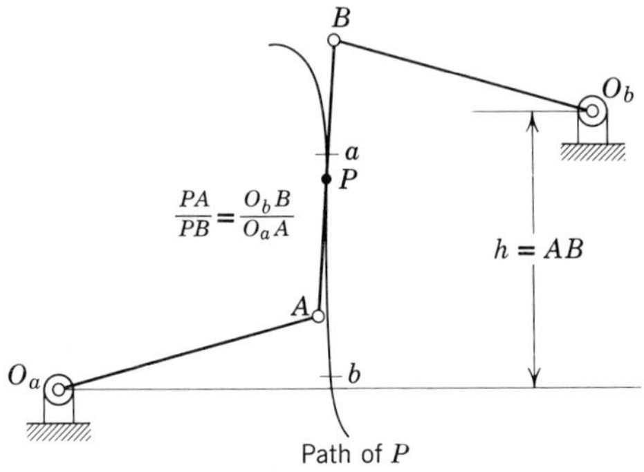

Fig. 5-32. Watt's straight-line linkage.

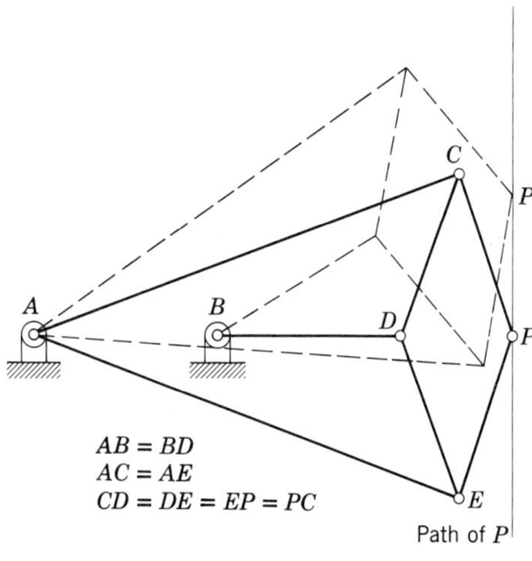

Fig. 5-33. The Peaucellier cell.

Point P on the coupler of a four-bar mechanism describes a path which has a straight portion. Part of the path of point P has been plotted in Fig. 5-32. Segment ab is approximately straight. The Peaucellier cell, shown in two positions in Fig. 5-33, generates an exact straight-line path.

Intermittent motion mechanisms. The problem of obtaining an intermittent motion is of concern to designers of automatic machinery. A survey of intermittent motion mechanisms has been made in an article by

Bogardus (Ref. 2). The article includes a good bibliography which will guide the reader interested in a more thorough study of these mechanisms.

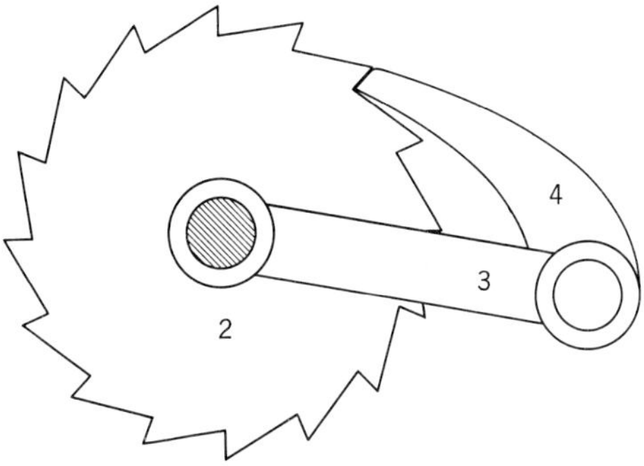

Fig. 5-34. Ratchet wheel and pawl.

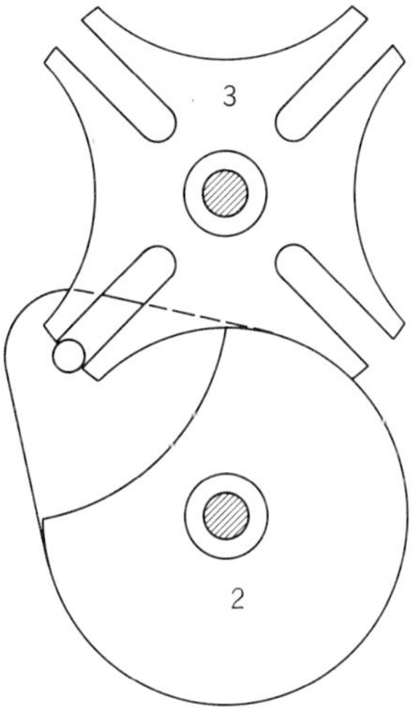

Fig. 5-35. Four-station external Geneva mechanism.

In the simple ratchet wheel and pawl mechanism of Fig. 5-34, a movement of the wheel in a counterclockwise direction is caused by a counterclockwise movement of link 3. When link 3 returns to its original position, the wheel remains at rest, the pawl, link 4, sliding over the teeth.

A mechanism widely used for indexing is the Geneva mechanism illustrated in Fig. 5-35. This is a four-station external Geneva. During a portion of each revolution of the driver, link 2, the slotted wheel, is turned through a 90° angle. During the remaining portion of the driver's revolution, the slotted wheel remains at rest, locked in place by the driver.

Fig. 5-36. Six-station external Geneva mechanism. (Courtesy Commercial Cam and Machine Company, Chicago, Ill.)

A six-station Geneva is shown in Fig. 5-36. Here the slotted wheel is moved 60° for each driver revolution.

Computing mechanisms. Computing mechanisms are surveyed in an article by Michalec (Ref. 9). Computing mechanisms are capable of doing a variety of mathematical operations, and they serve many functions in mechanical units controlling processes.

The simple linkage differential of Fig. 5-37 can add or subtract. For example, if the movement of the upper slider represents some quantity x, and that of the lower slider y, then the center slider's movement will approximately equal half the sum of x and y.

The Scotch yoke mechanism can be employed, as in Fig. 5-38, as a sine or cosine generator. The movement of the slide is proportional to the sine of θ or of the cosine of θ'.

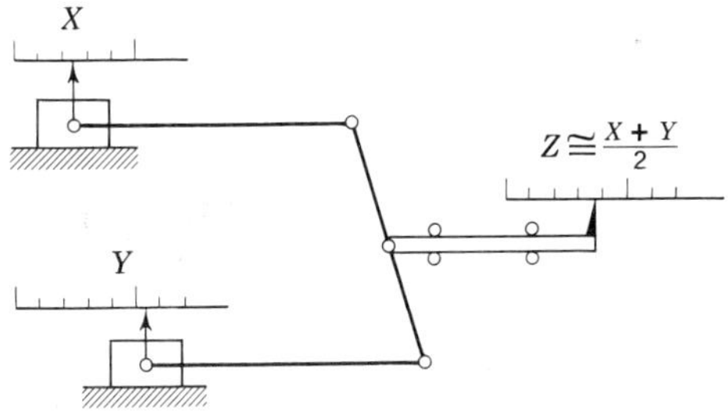

Fig. 5-37. Simple linkage differential.

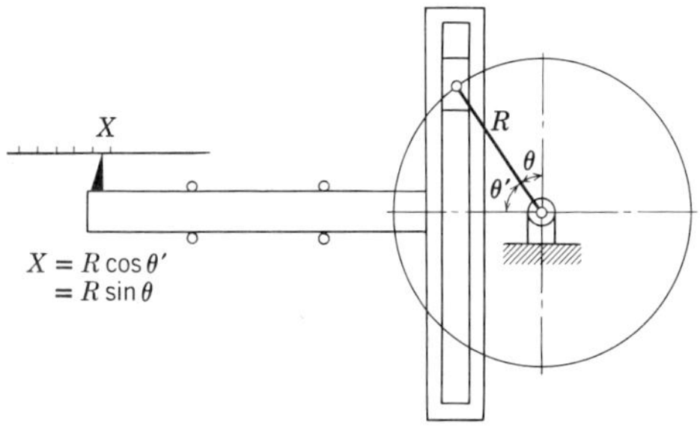

Fig. 5-38. Scotch yoke used as a sine-cosine generator.

REFERENCES

1. J. S. Beggs, *Mechanisms*, ch. 13, McGraw-Hill Book Co., New York, 1955.
2. F. J. Bogardus, "A survey of intermittent mechanisms," *Machine Design*, September 20, 1959.
3. G. A. Burls, "Note on maximum crosshead velocity," *Proceedings of the Institution of Civil Engineers*, v. 131 (1898), pp. 338-346.
4. Chin-U Ip and L. C. Price, "A simple formula for determining the position of maximum slider velocity in a slider-crank mechanism," Trans. ASME, v. 80 (1958), pp. 415–418.
5. Bruce L.Harding, "Standardized notation for and classification of four-bar linkages," *Machine Design*, November 28, 1957.
6. H. L. Horton, *Ingenious Mechanisms for Designers and Inventors*, v. III, The Industrial Press, New York.
7. John A. Hrones and George L. Nelson, *Analysis of the Four-Bar Linkage*, published jointly by The Technology Press of M.I.T., Cambridge, Mass., and John Wiley and Sons, New York, 1951.
8. F. D. Jones, *Ingenious Mechanisms for Designers and Inventors*, vs. I and II, The Industrial Press, New York.

9. George W. Michalec, "Analog computing mechanisms," *Machine Design*, March 19, 1959.
10. Sigmund Rappaport, "5 Linkages for Straight-line Motion," *Product Engineering*, October 12, 1959.
11. W. C. Unwin, "Determination of crank angle for greatest piston velocity," *Proceedings of the Institution of Civil Engineers*, v. 125 (1896), pp. 363–366.

PROBLEMS

5-1. A four-bar mechanism is shown in Fig. P5-1. For each set of link proportions in the table below, determine the type of mechanism described.

Mechanism No.	O_aA	AB	O_bB	O_aO_b
1	5.0	1.4	4.2	1.0
2	2.5	2.0	1.8	1.1
3	1.0	2.7	1.8	3.0
4	1.0	4.9	3.0	3.9
5	2.5	1.1	1.8	2.0
6	2.7	3.0	1.8	1.0
7	1.1	8.5	2.3	1.9

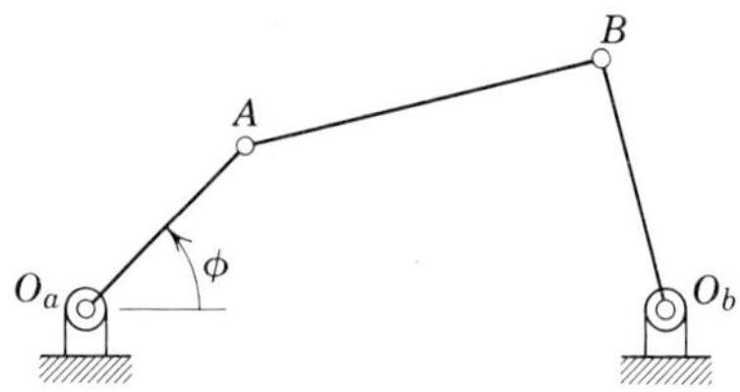

Fig. P5-1

5-2. In a four-bar mechanism (Fig. P5-1) the link lengths are: $O_aA = 2.0$, $AB = 5.4$, $O_bB = 3.6$, and $O_aO_b = 6.0$. The fixed link is O_aO_b. The driving link, O_aA, turns at a constant angular velocity of 120 rpm ccw. Find the angular velocity of driven link, O_bB, at the phases $\phi = 0°, 45°, 130°, 180°, 225°$, and $300°$.

5-3. Find the mechanical advantage of the mechanism of Prob. 5-2 for each of the specified phases.

5-4. In a four-bar mechanism (Fig. P5-1) the link lengths are: $O_aA = 1.3$, $AB = 2.1$, $O_bB = 2.3$, and $O_aO_b = 2.9$. The fixed link is O_aO_b and O_bB is the driver. At what driver positions do dead points occur?

For Probs. **5-5** to **5-9** the origin of the coordinate axes is to be placed near the lower left corner of a sheet of $8\frac{1}{2} \times 11$ paper held with the long dimension vertical.

5-5. Link AB (Fig. P5-5) moves from A_1B_1 to A_2B_2. Find the virtual pole for this movement.

5-6. Link AB (Fig. P5-6) moves from A_1B_1 to A_2B_2. Find the virtual pole for this movement.

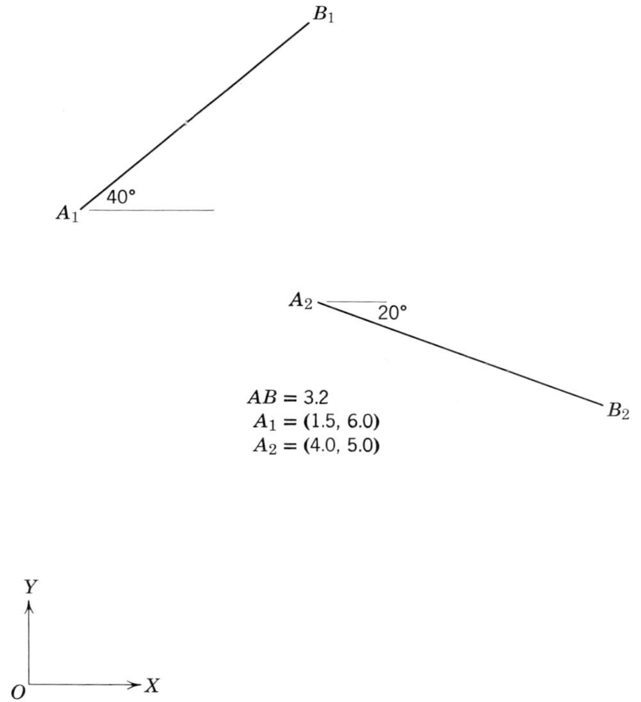

Fig. P5-5

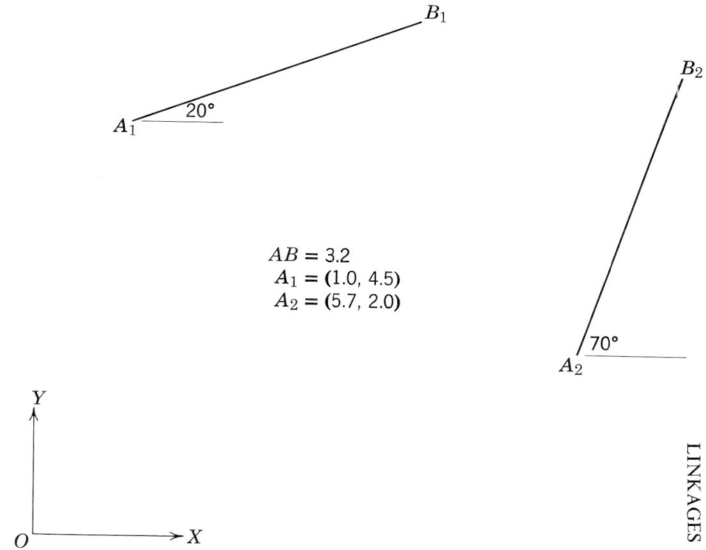

Fig. P5-6

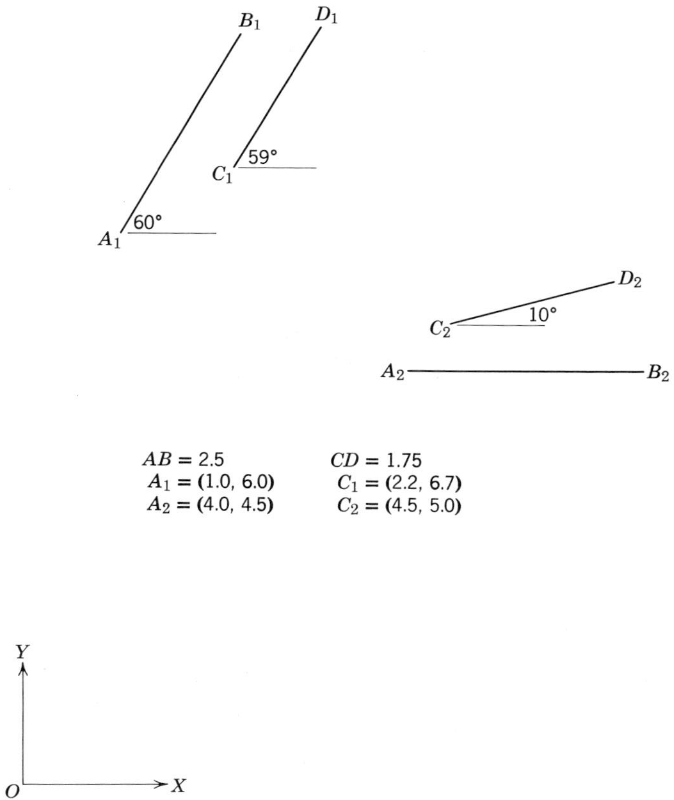

Fig. P5-7

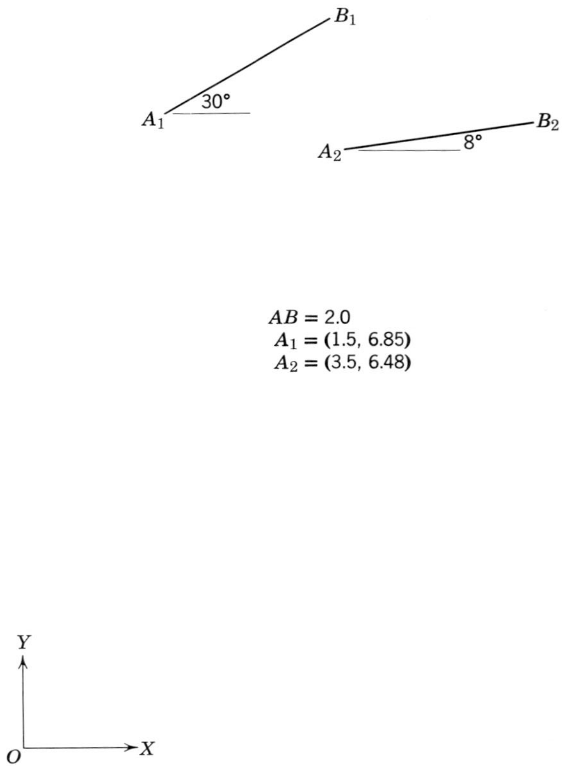

Fig. P5-8

5-7. While AB (Fig. P5-7) moves from A_1B_1 to A_2B_2, CD moves from C_1D_1 to C_2D_2. Find the final position of CD *relative to* an observer on AB stationed at A_1B_1. Construct the relative pole for the movement of CD relative to AB.

5-8. In Fig. P5-8 A_1B_1 and A_2B_2 are two positions of the coupler of a four-bar mechanism. Design a four-bar mechanism which will carry this link through the two prescribed positions. Determine the mechanical advantage at each of the two positions for your mechanism, assuming O_aA is the driver.

5-9. In Fig. P5-9 A_1B_1, A_2B_2, and A_3B_3 are three positions of a link. Design a four-bar mechanism of which AB is the coupler.

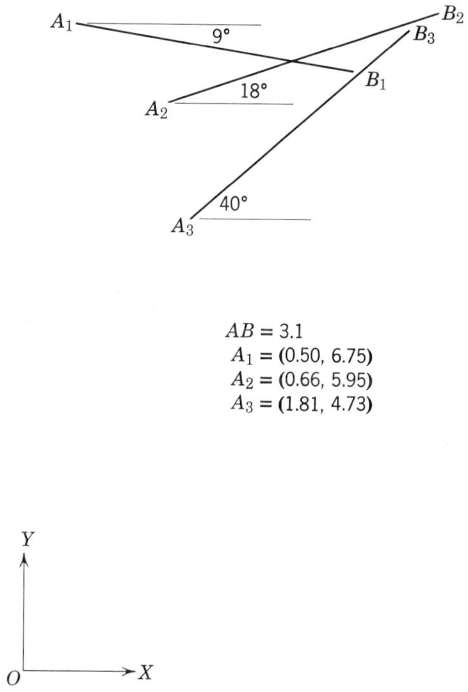

Fig. P5-9

5-10. Design a four-bar mechanism in which a 32° cw rotation of the driving link is accompanied by a 17° cw movement of the driven link.

5-11. Design a four-bar mechanism in which a 68° ccw movement of the driver is accompanied by a 29° cw movement of the driven link.

5-12. A four-bar mechanism is used in a process control mechanism to coordinate movements of two links. Design a mechanism which will conform to the following schedule of angular displacements.

Driving link	*Driven link*
$\Delta\phi_{12} = 27°$ cw	$\Delta\psi_{12} = 15°$ cw
$\Delta\phi_{23} = 22°$ cw	$\Delta\psi_{23} = 11°$ cw

5-13. Figure P5-13 shows a four-bar mechanism. Draw the mechanism at the $\phi = 45°$ phase and then plot the path of point P on the coupler. Assume O_aA is the driver and plot for each $10°$ of driver rotation.

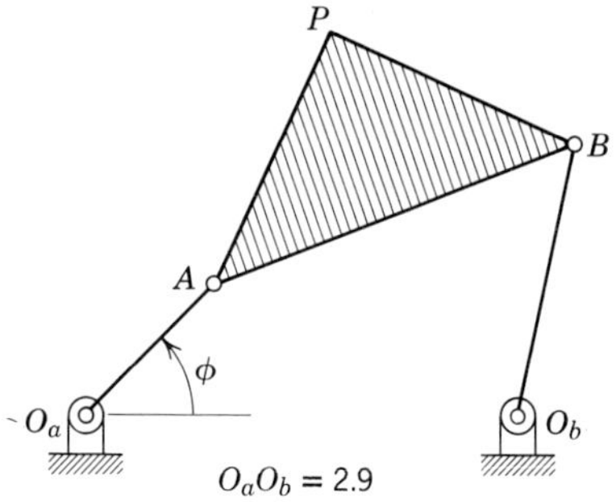

$O_aO_b = 2.9$
$O_aA = 1.2$
$AB = 2.6$
$O_bB = 1.8$
$AP = 1.8$
$BP = 1.8$

Fig. P5-13

5-14. A quick-return mechanism is shown in Fig. P5-14. The crank, O_bB, turns at a constant velocity. Determine the stroke of the piston and the time ratio of forward stroke to return stroke.

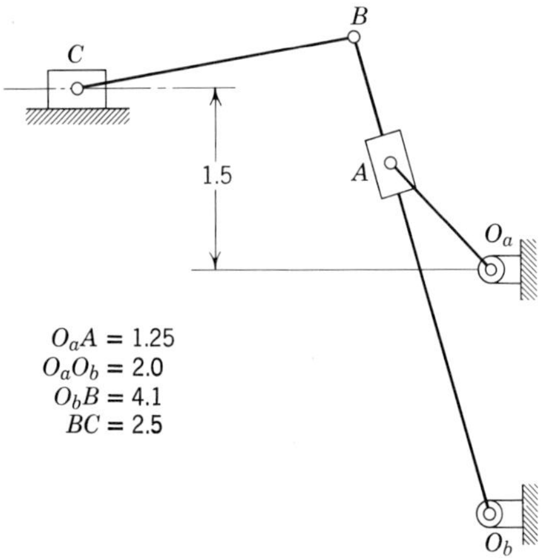

$O_aA = 1.25$
$O_aO_b = 2.0$
$O_bB = 4.1$
$BC = 2.5$

Fig. P5-14

5-15. An in-line slider mechanism (Fig. 5-22*a*) has a crank length of 4 in. and a connecting-rod length of 16 in. The crank turns ccw at 720 rpm. Lay out the mechanism at the $\phi = 45°$ phase and determine:
a. the slider displacement at each $15°$ crank position,
b. the slider velocity at each $15°$ crank displacement,

c. plot a curve for slider displacement versus crank angle,
d. plot a curve for slider velocity versus crank angle,
e. plot a velocity-space curve for the slider, plotting the slider velocity as ordinate at the corresponding slider position.

5-16. For the mechanism of Prob. 5-15, determine the connecting rod's angular velocity at $\phi = 0°, 45°, 135°$, and $180°$, using information derived from graphical constructions.

5-17. For the mechanism of Prob. 5-15, assume the crank is the driver and determine the mechanical advantage at $\phi = 0°, 45°, 75°, 135°$, and $180°$.

5-18. For the mechanism of Prob. 5-15, determine graphically the slider acceleration at $\phi = 0°, 45°, 75°, 135°$, and $180°$.

5-19. For the mechanism of Prob, 5-15, calculate the slider displacement, velocity, and acceleration at $\phi = 0°, 45°, 75°, 135°$, and $180°$.

5-20. For the mechanism of Prob. 5-15, calculate the angular velocity of the connecting rod at $\phi = 0°, 45°, 75°, 135°$, and $180°$.

5-21. For the mechanism of Prob. 5-15, calculate the angular acceleration of the connecting rod at $\phi = 0°, 45°, 75°, 135°$, and $180°$.

5-22. For the mechanism of Prob. 5-15, determine the crank positions at which the maximum slider velocity will occur.

5-23. Design an in-line slider-crank mechanism in which a crank movement of 38° cw will be accompanied by a slider movement of 1.2 in. to the right. At its initial position the slider is 3.4 in. to the right of the crank center.

5-24. A Watt straight-line mechanism is shown in Fig. P5-24. Draw the mechanism at the phase shown and plot the path of point *P*. Mark and measure the straight line segment of the path of *P*.

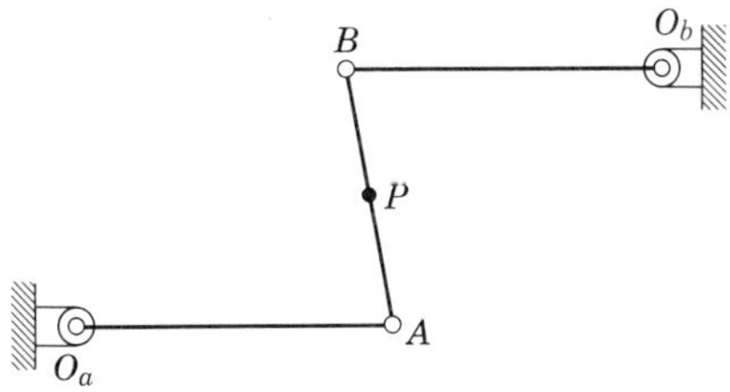

$O_aA = O_bB = 2.0$
$AP = 1/2\ AB$
O_b is 3.6 to right and 1.5 above O_a

Fig. P5-24

6 Cams

6-1. Introduction

Rothbart (Ref. 2) has defined a cam in the following way: "A cam is a mechanical member for transmitting a desired motion to a follower by direct contact." Cam mechanisms are versatile members of the mechanism family. They can produce complicated output motions difficult to achieve with other types of mechanisms. They find use widely in automatic machines, machine tools, internal combustion engines, control mechanisms and many other applications.

There are at least three members in a cam mechanism: the cam, the follower, and the frame. Almost always the cam is the driver and the follower the driven.

Cam mechanisms may be grouped into two broad categories: two dimensional (2-D) and three dimensional (3-D). The distinction is illustrated in Fig. 6-1. For the 2-D cam mechanism in Fig. 6-1*a*, the output displacement, z, of the follower is determined by a single input displacement, y, the angular movement of the cam. On the other hand, for the 3-D mechanism in Fig. 6-1*b*, the radial displacement of the follower, z, is determined both by the angular movement, y, of the cam and the lateral movement, x, of the follower.

6-2. Cam Types

Cams may be classified in a number of ways: by their shape, by the manner of constraint of the follower, and by the general type of motion imparted to the follower.

Three typical 3-D cams are shown in Fig. 6-2, and some of the most widely used 2-D cams are illustrated in Fig. 6-3. The face, radial, and drum cams have rotary motion, while the bar cam has a motion of translation.

A roller follower operating in the groove of either the face or the drum cam will be positively constrained: it cannot go out of contact with the

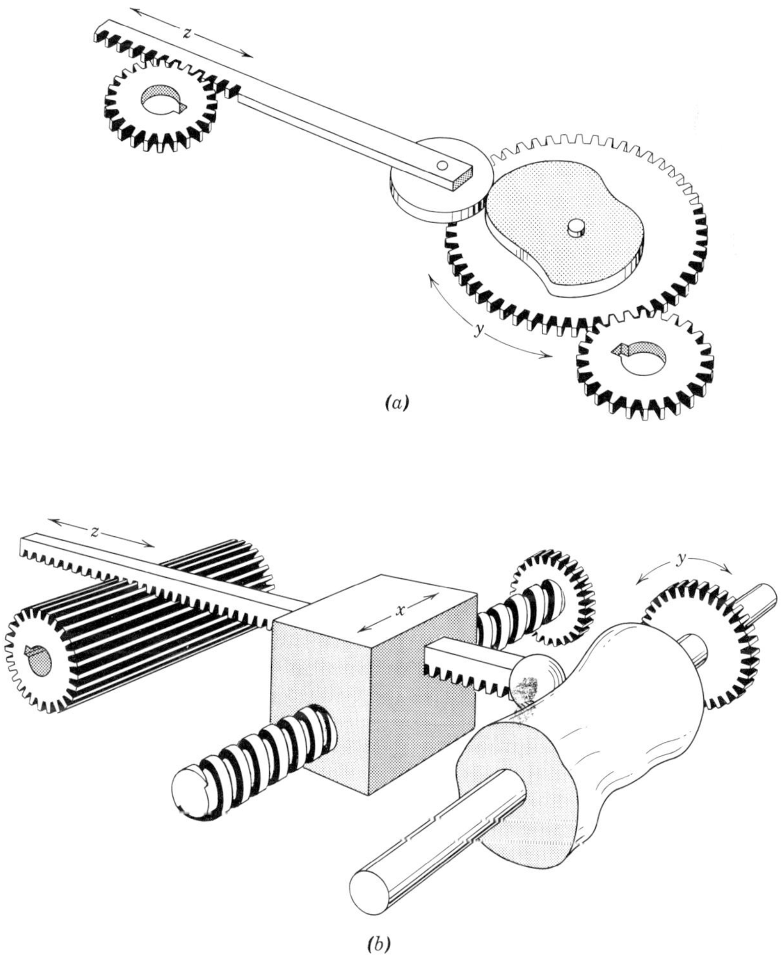

Fig. 6-1. Cam mechanisms. (*a*) 2-D. (*b*) 3-D. (Courtesy The Parker-Hartford Corporation, Hartford, Conn.)

cam in normal operation. But for the radial or bar cams, some external means, usually a spring, is needed to maintain contact between the follower and the cam.

The three basic types of follower motion pattern imparted to the follower by the cam are illustrated in Fig. 6-4. They are distinguishable from each other by the presence or absence of dwells and the positioning of the dwells. In each diagram the follower displacement is s and the cam displacement θ. A cam which causes the follower to rise, return, and

Fig. 6-2. Typical three-dimensional cams. (Courtesy The Parker-Hartford Corporation, Hartford, Conn.)

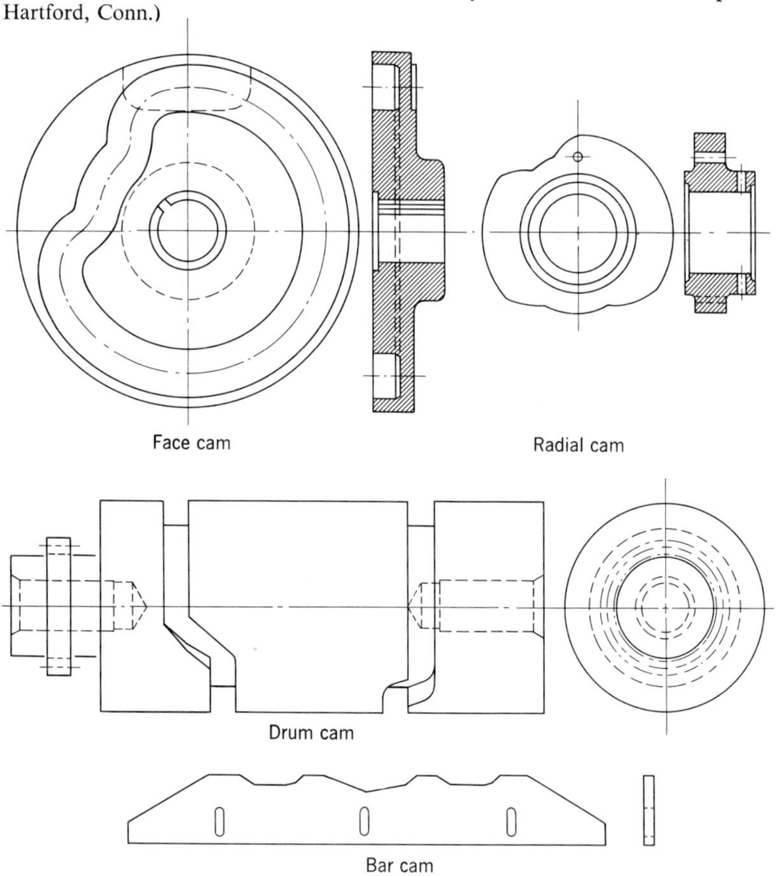

Fig. 6-3. Two-dimensional cam types. (Courtesy The Cincinnati Milling Machine Company, Cincinnati, Ohio.)

immediately rise again with no dwell is a rise-return-rise (R-R-R) cam (Fig. 6-4*a*). This is the least used type in industry. The dwell-rise-return-dwell (D-R-R-D) type, Fig. 6-4*b*, is used more frequently. And the most widely used type is the dwell-rise-dwell-fall-dwell (D-R-D-F-D) of Fig. 6-4*c*. Usually the name of this type is abbreviated to dwell-rise-dwell

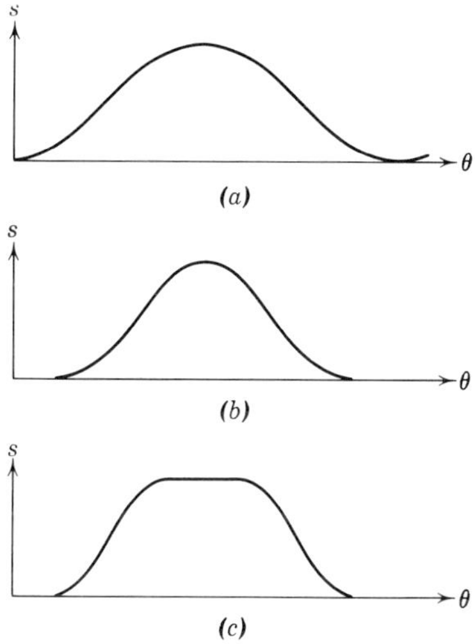

Fig. 6-4. Follower motion patterns. (*a*) Rise-return-rise (R-R-R). (*b*) Dwell-rise-return-dwell (D-R-R-D). (*c*) Dwell-rise-dwell-fall-dwell (D-R-D-F-D).

(D-R-D), for the analysis of the dwell-fall-dwell portion of the cam is similar to that for the dwell-rise-dwell portion.

6-3. Follower Types

Cam followers may be classified by their shape, movement, and the location of their line of movement. A number of followers are shown in Fig. 6-5.

Three of the common shapes are illustrated: knife-edge, roller, and flat-face. The roller and flat-face profiles are those most frequently used.

In Fig. 6-5*a* and 6-5*c* the line of movement of the follower passes through the center of rotation of the cam. These are in-line, or radial, followers. The line of movement of the roller follower in Fig. 6-5*b* is offset from the cam's center of rotation. It is thus an offset follower.

The motion of the follower will be either reciprocating or oscillating.

An oscillating follower is used in the cam design of Fig. 6-5*d*. The remaining followers are reciprocating.

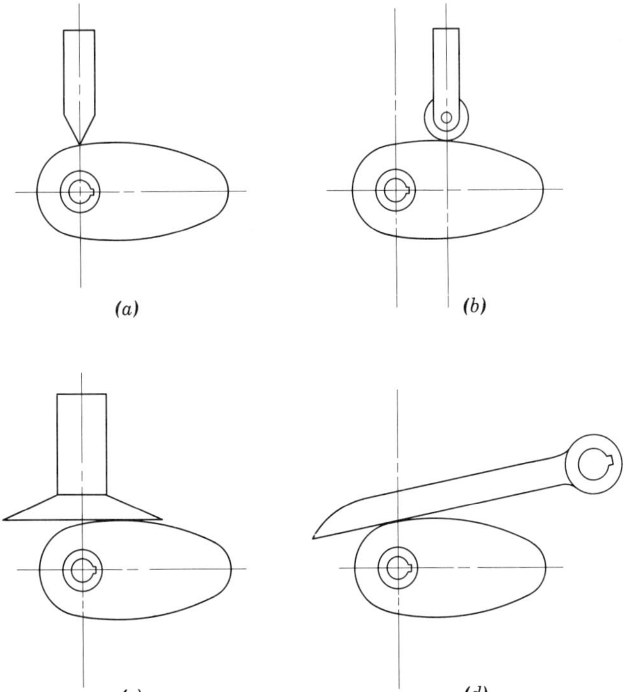

Fig. 6-5. Follower types. (*a*) Knife-edge, in-line. (*b*) Roller, offset. (*c*) Flat-face-in-line. (*d*) Oscillating.

6-4. Follower Displacement Diagrams

The remainder of this chapter will be devoted to a discussion of the design of radial cams. Two fundamental graphical constructions will be involved in each cam design: (1) the construction of the follower displacement diagram, a graph defining follower movement; and (2) the layout of the cam profile itself.

First the problem of constructing follower displacement diagrams will be considered. A typical diagram is shown in Fig. 6-6. It is a graph in which follower displacement has been plotted against cam displacement. Cam displacement is plotted to a convenient scale from 0° to 360°, since follower motion repeats itself each cam revolution. The follower displacement is measured from the lowest follower position and plotted to whatever scale will be used in making the cam profile layout. For a reciprocating follower, the follower displacement is a linear distance. For an oscillating follower, either of two methods could be used: the angular displacement of the follower could be plotted, or the linear displacement of some reference point on the follower along a circular arc could be plotted.

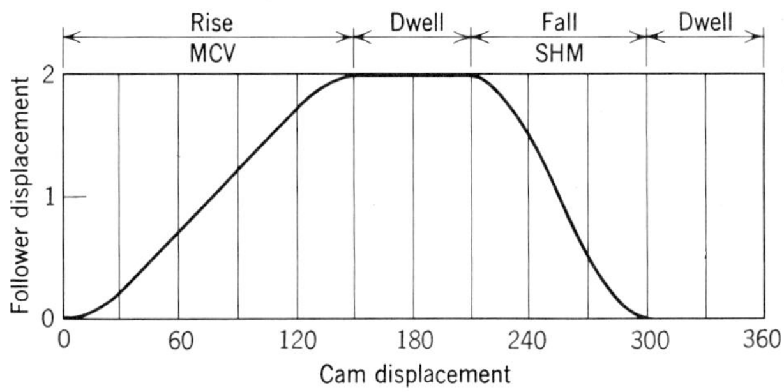

Fig. 6-6. Typical follower displacement diagram.

In the next few articles, some of the common follower motions will be investigated.

6-5. Simple Harmonic Motion

This has been a popular follower motion. It is easy to lay out and easy to calculate, using the equations of Art. 2-14. Since it is easy to determine the follower acceleration at each cam position, the problem of determining inertia forces is simplified.

Construction. Figure 6-7 has two examples of the construction. In Fig. 6-7*a* the follower is to rise 2 in. while the cam turns from 0° to 180°, a cam displacement interval of 180°. In the second example (Fig. 6-7*b*), the follower is to fall 2 in. while the cam turns from 210° to 330°, a cam displacement interval of 120°. The following construction was employed to obtain the follower displacement curves.

1. Divide the cam displacement interval into n equal divisions, where n is usually an even number. In each example of Fig. 6-7 six divisions were used. The ordinates were numbered from 0 to 6.

2. Construct a semicircle having the cam rise (or fall) as diameter. Divide this semicircle into n equal arcs. This is the *harmonic circle*. In the examples six equal arcs were used, since in each case the cam displacement

interval had been divided into six divisions. The intercepts were numbered from 0 to 6, from bottom to top for the first example (rise) and top to bottom in the second (fall).

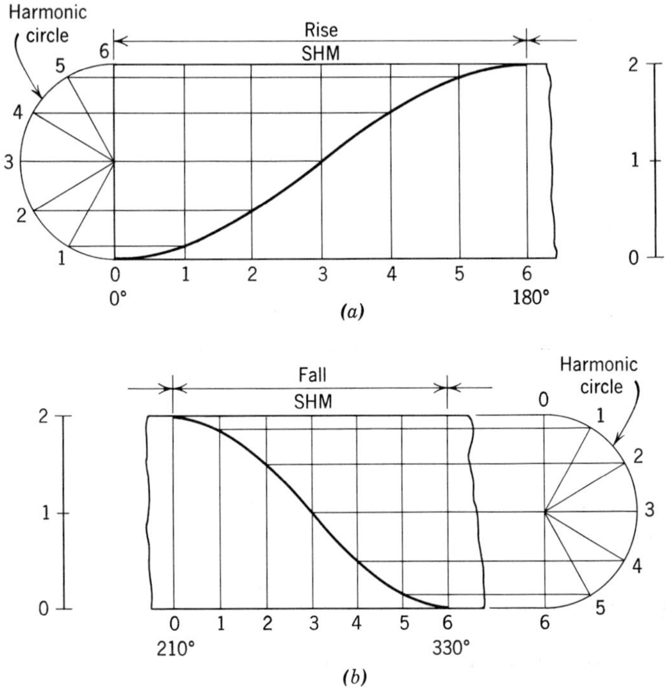

Fig. 6-7. Follower displacement diagrams: SHM.

3. Project the intercepts from the harmonic circle to the corresponding ordinates, 1 to 1, 2 to 2, etc.
4. Connect these points with a french curve to obtain the curve.

Discussion. Figure 6-8 shows typical displacement, velocity, and acceleration curves for a follower rising with simple harmonic motion. The acceleration curve has been plotted with the assumption that the rise is preceded and followed by dwells. While the velocity varies smoothly during the rise, the acceleration curve shows sudden changes in acceleration at the beginning and end of the curve. This is highly undesirable for cams moving at high speeds, because noise and vibration will result. Consequently, SHM should be used only at low and medium speeds.

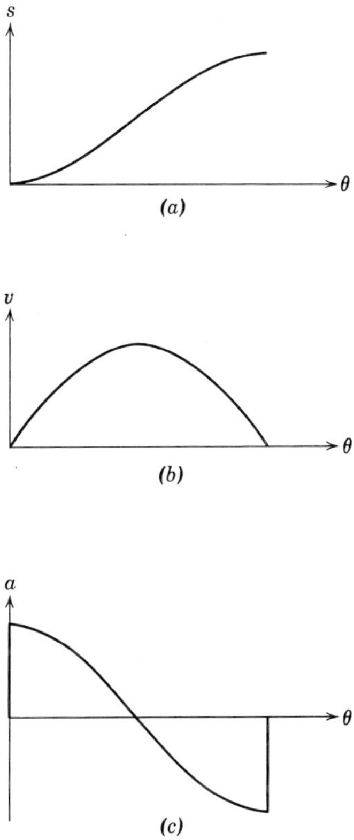

Fig. 6-8. Characteristics of SHM follower motion.

6-6. Constant Acceleration

Like the SHM follower motion, constant acceleration has the advantage of being easy to lay out and to calculate. The equation for rectilinear motion with constant acceleration is

$$s = v_o t + \tfrac{1}{2}at^2 \tag{2-3}$$

At the instant the follower begins its rise or fall, its velocity is zero, $v_o = 0$. Therefore the equation of its motion becomes

$$s = \tfrac{1}{2}at^2 \tag{6-1}$$

If the cam is rotating at constant velocity,

$$\theta = \omega t \tag{2-9a}$$

or

$$t = \frac{\theta}{\omega}$$

Substituting in Eq. 6-1,

$$s = \frac{a}{2\omega^2}\theta^2 \tag{6-2}$$

Since a and ω are constant,

$$s \propto \theta^2 \tag{6-3}$$

The fact that the follower displacement is directly proportional to the square of the cam displacement justifies the following construction.

Construction. An example of the construction for a constant acceleration follower curve is shown in Fig. 6-9. The follower is to rise 2 in. while the

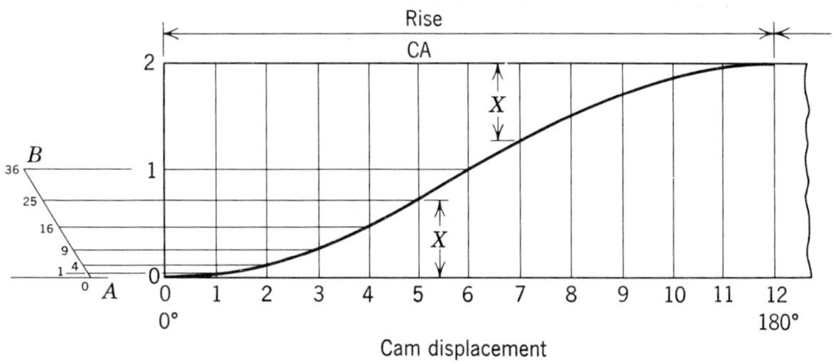

Fig. 6-9. Follower displacement diagram: constant acceleration.

cam has an angular displacement of 180°. The curve is constructed so that the follower rises halfway with constant acceleration, then rises the remaining distance with constant deceleration.

1. Divide each half of the cam displacement interval into n equal divisions. In the example, each half of the cam displacement interval has been divided into six parts, or 12 for the whole interval.
2. Divide half the follower rise into n^2 equal divisions. This can be done in a number of ways. In the example, some length was selected on a rule such that it had 36 divisions (6^2) and the rule placed so that the length reached between lines from the beginning and the midpoint of the follower rise. This is line AB.
3. Project the end of the first of the n^2 increments to the first ordinate, the end of the fourth increment to the second ordinate, the end of the ninth increment to the third ordinate, etc.
4. The second half of the curve is similar to the first and can be constructed by transferring the appropriate distances with dividers. In the example, distance x on ordinate 5 and distance x on ordinate 7 correspond.

There is a correspondence between ordinates 4, and 8, 3 and 9, etc.

5. Connect the points with a french curve.

Discussion. The characteristics of the constant acceleration follower curve are illustrated in Fig. 6-10. The acceleration curve has abrupt

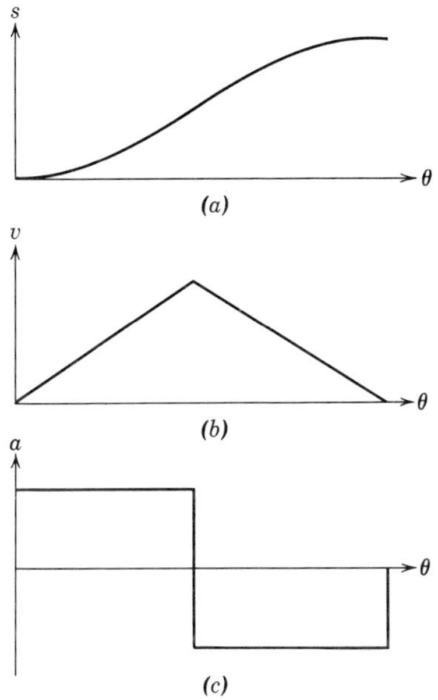

Fig. 6-10. Characteristics of the constant acceleration follower curve.

changes at the beginning, transition point, and end of the follower's travel. These abrupt changes in acceleration are detrimental to the operation of the machine. Hence this type of curve should be applied only at low speeds.

6-7. Modified Constant Velocity

The simplest follower motion would involve a rise (or fall) at constant velocity. The characteristics of a constant velocity follower motion are illustrated in Fig. 6-11. It has been assumed, in constructing these curves, that the rise was preceded and followed by dwells. At the beginning and end of the rise there are sudden changes in velocity (Fig. 6-11*b*), changes that occur in zero time. This implies, as Fig. 6-11*c* shows, that the accelerations are infinite in magnitude, a physical impossibility.

To avoid sudden velocity changes, while retaining substantially a constant velocity motion, constant acceleration intervals are introduced at each end of the follower motion. The follower accelerates uniformly from rest to some velocity, v, rises further at constant velocity, then completes its rise while decelerating from v to rest.

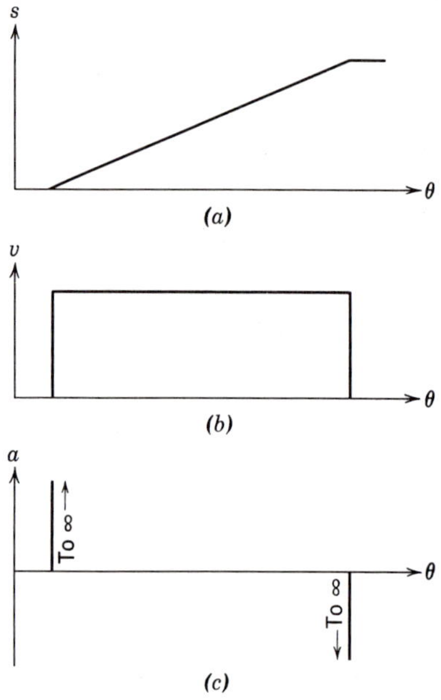

Fig. 6-11. Characteristics of constant velocity follower motion.

Construction. An example of the construction is in Fig. 6-12*a*. The follower rises 2 in. while the cam turns 120°.

1. Divide the cam displacement into three intervals: constant acceleration, constant velocity, and constant deceleration. In the example, the intervals were made 30°, 90°, and 30°.

2. Locate points A and B, as shown, at the midpoints of the first and third intervals. Placing a straight edge on points A and B, locate the constant velocity segment, CD, of the follower curve.

3. The constant acceleration and deceleration segments of the follower curve are then filled in using the method of the preceding article.

Proof. The key to the above method is the location of points A and B. It needs to be proved that they are in fact at the locations where the straight line segment of the follower curve, if extended, would intercept the bottom and top lines of the graph. An enlargement of the lower left portion of the graph is shown in Fig. 6-12*b*.

1. Regard curve EC and line EAC as two possible follower curves. Then it must be proved that A is at the midpoint of the constant acceleration interval, that is, that

$$t_2 = \tfrac{1}{2}t_1$$

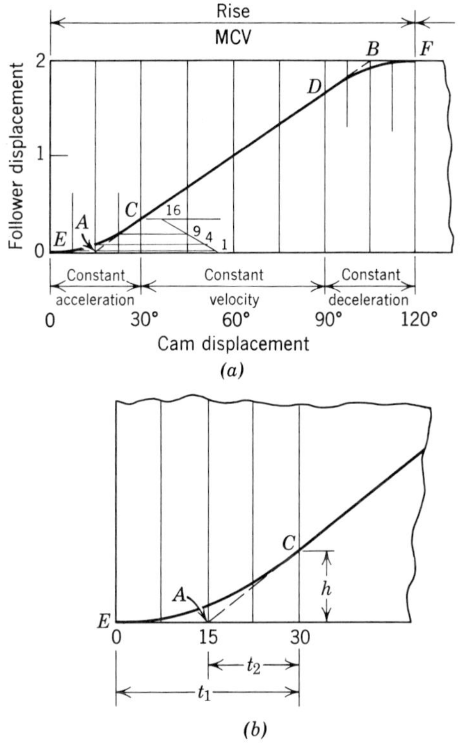

Fig. 6-12. Follower displacement diagram: modified constant velocity.

2. Curve EC is a constant acceleration curve; hence the equations for constant acceleration rectilinear motion apply. Thus,

$$h = \tfrac{1}{2}at_1^2$$

and $$v = at_1$$

Then $$h = \tfrac{1}{2}(at_1)t_1 = \tfrac{1}{2}vt_1$$

3. For the follower curve, EAC, the follower rests from E to A and then rises, A to C, a distance h at velocity v in time t_2. Therefore,

$$h = vt_2$$

4. Equating the two expressions for h,

$$vt_2 = \tfrac{1}{2}vt_1$$

or

$$t_2 = \tfrac{1}{2}t_1$$

which proves that A lies at the midpoint of the constant acceleration interval. A similar proof applies to point B.

Discussion. The characteristics of the modified constant velocity follower motion are illustrated in Fig. 6-13. The acceleration curve

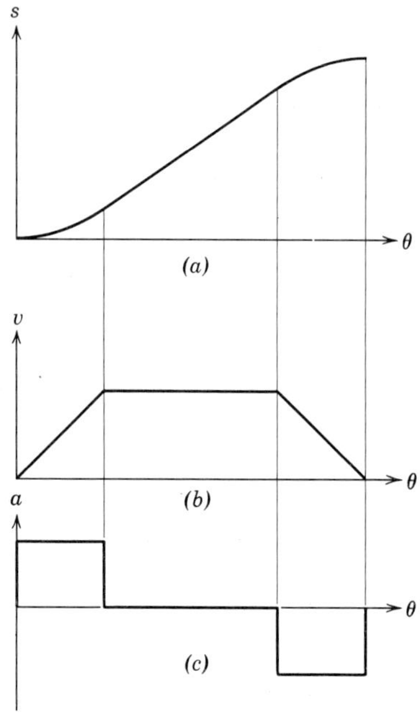

Fig. 6-13. Characteristics of modified constant velocity follower motion.

reveals sudden changes in acceleration at the beginning and end of each constant acceleration interval. These abrupt changes in acceleration will produce noise, vibration, and excessive wear at high speeds. This type of follower motion must be restricted to low speeds.

6-8. Cycloidal Motion

A cycloid is the path of a point on a circle made as the circle rolls on a line. In Fig. 6-14*a* a circle is imagined to roll on the vertical axis of the displacement diagram. By making the circumference of this circle equal to the desired follower rise, the vertical displacement of P in one revolution of the generating circle will be equal to the follower rise. A follower moving such that its vertical displacement always corresponds to that of P is said to have cycloidal motion.

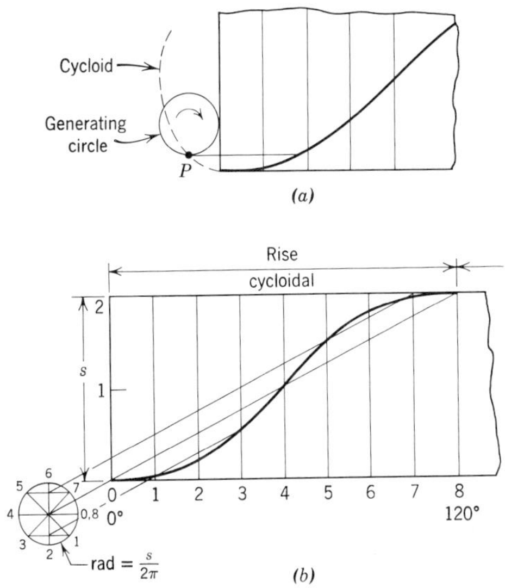

Fig. 6-14. Follower displacement diagram: cycloidal motion.

The following graphical procedure can be used to determine the follower curve.

Construction. An example is shown in Fig. 6-14*b*. The follower rises 2 in. while the cam turns 120°.

1. Divide the cam displacement interval into n parts, where n is an even number (for convenience). Number from 0 to n.
2. Draw the diagonal of the diagram, as shown in the example, extending it below the diagram.
3. Draw a circle of radius $S/2\pi$ having its center anywhere on the lower portion of the diagonal. Divide the circle into n equal arcs and number them as shown in the example.
4. Project the circle points to the circle's vertical diameter.
5. Project each of these points in a direction parallel to the diagonal to the corresponding ordinate.
6. Join the points with a french curve.

Discussion. The characteristic curves of Fig. 6-15 show that the cycloidal follower motion is the smoothest of those discussed. The acceleration has no discontinuities, varying smoothly in a sine-curve pattern. Cycloidal motion is recommended for high speeds.

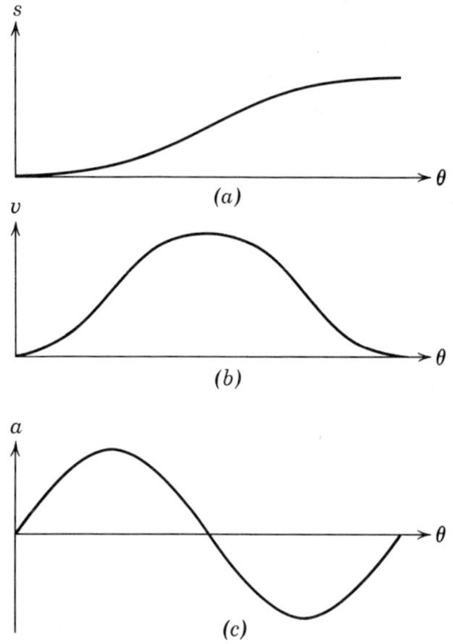

Fig. 6-15. Characteristics of cycloidal follower motion.

6-9. Cam Layout

The second of the graphical constructions needed in designing a cam mechanism is the layout of a cam profile which will impart the motion to the follower prescribed by the follower displacement diagram. In the actual profile layout, it is convenient to use the principle of inversion: the cam is imagined to be stationary and the follower moved around it. The layout of the cam must always, then, proceed in a direction opposite that of the cam's rotation, for the motion of the follower relative to the cam is opposite that of the cam relative to it.

The general approach to cam profile layout will be illustrated by the design of radial cams in the next few articles. The actual graphical techniques used in the examples are not the only ones that could be used. There is generally more than one graphical technique available.

6-10. Radial Cam—Translating Knife-Edge Follower

An example of the construction is in Fig. 6-16. The following steps are involved:

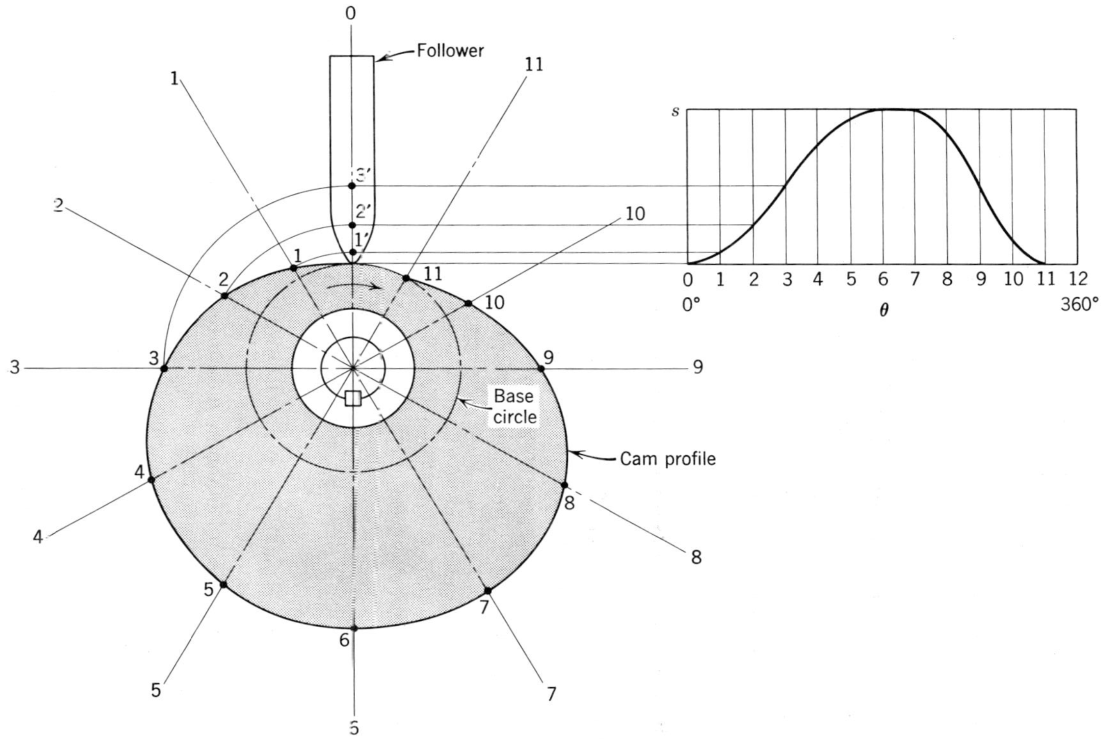

Fig. 6-16. Radial cam—knife-edge, translating, in-line follower.

1. Select a base circle diameter. Draw the base circle, and draw the follower in its lowest (0°) position. The base circle is the smallest circle drawn to the cam profile from the cam center. Its size is, within limits, arbitrary.

2. Draw the displacement diagram with its horizontal axis in line with the 0° position of the follower edge.

3. Draw radial lines dividing the cam into angular intervals corresponding to those used in the displacement diagram. Number in a direction opposite that of the cam's rotation.

4. Project points from the displacement curve to the center line of the follower. Point 1 projects to 1′, 2 to 2′, etc.

5. Setting a compass or divider point at the cam center, revolve point 1′ to radial line 1, 2′ to radial line 2, 3′ to 3, etc. These are the alternate positions of the follower edge.

6. Draw a smooth curve through these points.

6-11. Radial Cam—Translating Roller Follower

The knife-edge follower of the previous article can follow accurately the cam profile, but the excessive friction and wear likely to occur make it impractical for most applications. A more practical follower uses a roller, usually a small bearing, which minimizes friction and wear.

In-line follower. The example in Fig. 6-17 is based on the same displacement diagram and base circle as for the knife-edge follower in Fig. 6-16, permitting a comparison of the cam shapes for different followers. The following construction steps are necessary to develop the cam profile:

1. Lay out the base circle and the follower in its lowest (0°) position, tangent to the base circle.

2. The center of the roller, point *P*, will be the trace point: that is, its alternate positions will be plotted in the same manner as were those of the edge of the knife-edge follower in Fig. 6-16. Therefore, draw the follower displacement diagram with its horizontal axis aligned with the lowest (0°) position of point *P*.

3. Draw radial lines to divide the cam into angular intervals corresponding to those used in the displacement diagram. Number in a direction opposite that of the cam rotation.

4. Project points from the displacement curve to the follower axis, 1 to 1′, 2 to 2′, 3 to 3′, etc. These are the alternate positions of the roller center, *P*, along the follower axis.

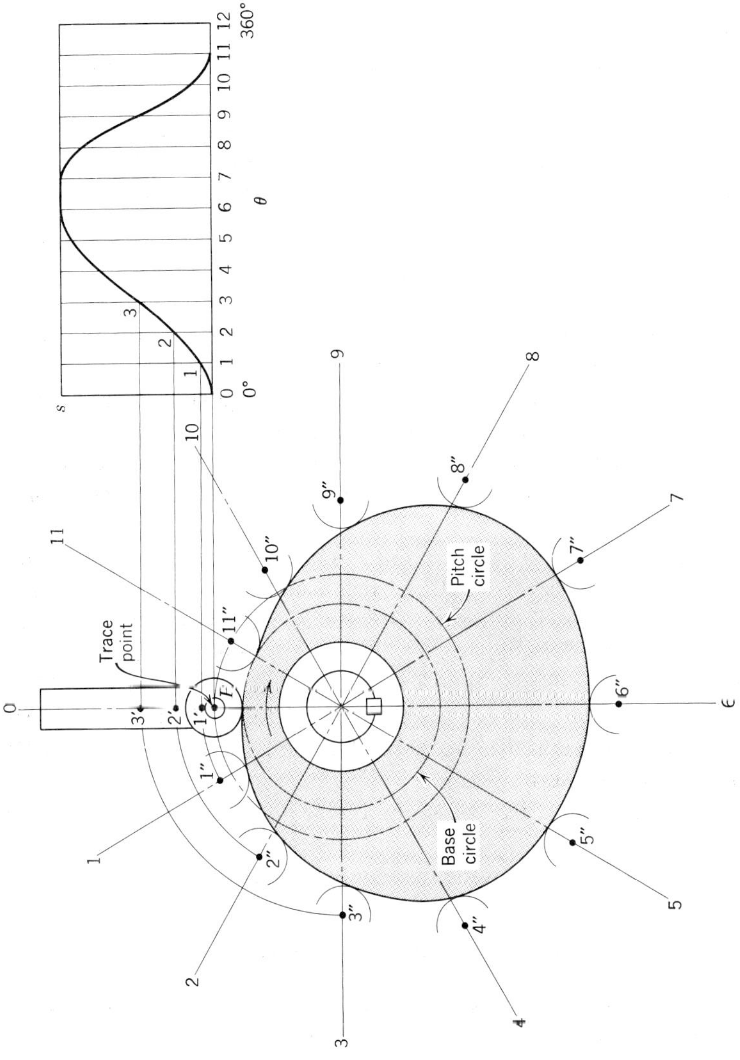

Fig. 6-17. Radial cam—roller, in-line, translating follower.

5. Next, the positions of point P relative to the cam must be plotted. Point $1'$ must be revolved 30°, point $2'$ must be revolved 60°, etc. Therefore, revolve each of these points to the corresponding radial line, $1'$ to $1''$, $2'$ to $2''$, $3'$ to $3''$, etc. Points $1''$, $2''$, $3''$, etc., are the positions of the roller center, P, relative to the cam.

6. Draw the roller profile at each of the relative positions.

7. Draw a smooth cam profile tangent to each of the roller profiles.

Offset follower. An example is shown in Fig. 6-18. The same base circle and displacement diagram are used as before, but this time the follower is offset to the left and the cam turns counterclockwise.

1. Lay out the base circle and the follower in its lowest (0°) position, tangent to the base circle. Again the roller center, P, will be the trace point.

2. Draw the follower displacement diagram with its horizontal axis aligned with the lowest (0°) position of the trace point, P.

3. Draw radial lines to divide the cam into angular intervals corresponding to those used in the follower displacement diagram.

4. Project points from the displacement curve to the follower axis, 1 to $1'$, 2 to $2'$, 3 to $3'$, etc. These are the alternate positions of the trace point along the follower axis.

5. The relative positions of the point P must be found as before. And, as before, $1'$ must be revolved 30°, $2'$ revolved 60°, etc. About the cam center, A, draw arc aa' of radius $A1'$ from radial line 0 to line 1. Draw arc bb' of radius $A2'$ from radial line 0 to line 2. Continue for each point.

6. Set dividers to offset $a1'$ and locate $1''$ by setting off from a' ($a'1'' = a1'$). Set dividers to offset $b2'$ and locate $2''$ by setting off from b'. Repeat for each point. Note that the directions of $a1'$ and $a'1''$ must be consistent; the same is true for those of $b2'$ and $b'2''$, etc.

7. At each of the relative positions of the roller center, P, draw the roller profile.

8. Draw a smooth cam profile tangent to each of the roller profiles.

6-12. Radial Cam—Translating Flat-Face Follower

The example of this construction, using the same displacement diagram and base circle as before, is shown in Fig. 6-19.

1. Draw the base circle and the follower in its lowest (0°) position, tangent to the base circle.

2. Construct the follower displacement diagram with its horizontal axis aligned with the lowest position of the follower face.

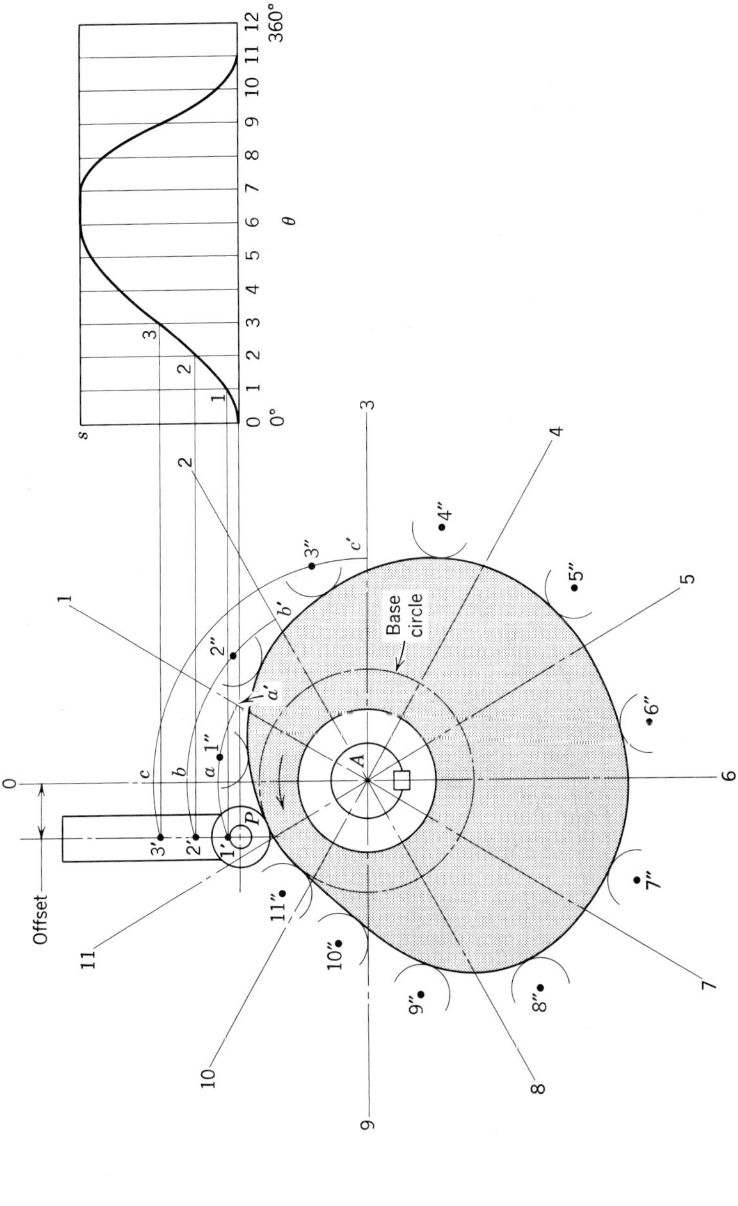

Fig. 6-18. Radial cam—translating, in-line, roller follower.

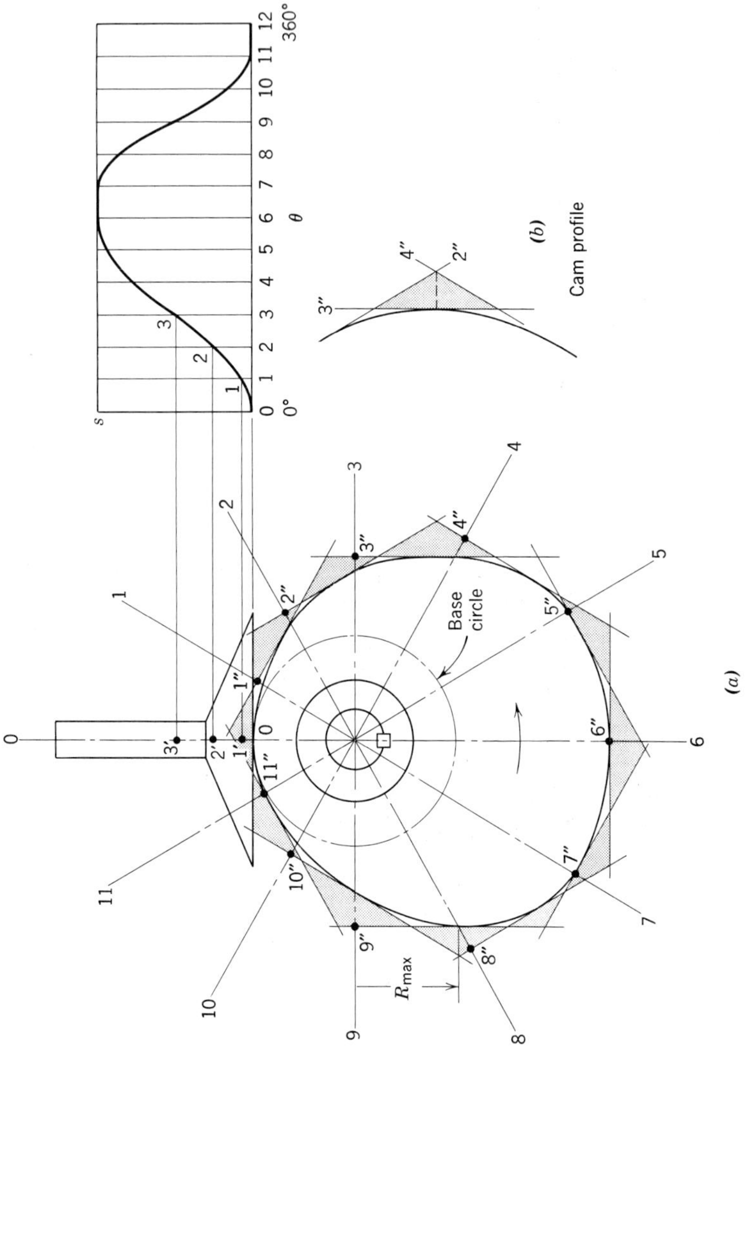

Fig. 6-19. Radial cam—translating, flat-face follower.

3. Draw radial lines to divide the cam into angular intervals corresponding to those used in the displacement diagram. Number in a direction opposite that of the cam's rotation.

4. Project points from the displacement curve to the follower axis, 1 to 1′, 2 to 2′, 3 to 3′, etc.

5. The positions of these points relative to the cam must be found. Revolve each point about the cam center to the corresponding radial line, 1′ to 1, 2′ to 2, etc. Points 1″, 2″, 3″, etc., thus found are the relative positions of the follower face.

6. Draw the follower profile at each position—in this case simply a line perpendicular to the appropriate radial line.

7. Draw a smooth cam profile tangent to each of these follower profiles. A rule illustrated in Fig. 6-19*b* is useful in guiding the eye while constructing the profile. When the follower profiles are drawn at each position, a lattice of triangles is formed. At each position the base of the triangle is the follower face for that position, and the remaining sides are the follower faces at the two adjacent positions. For example, in Fig. 6-19*b*, the base is the follower profile for position 3; the other sides are the follower profiles for positions 2 and 4. The tangency point between the cam profile and the follower profile at position 3 is *roughly* at a point directly under the outer vertex of the triangle, as Fig. 6-19*b* illustrates. The same rule applies to each of the other positions.

6-13. Radial Cam—Oscillating Roller Follower

An example of the construction is given in Fig. 6-20. It is assumed that the base circle diameter, follower pivot position, follower radius, roller radius, and angular travel of follower were either given or have been selected.

1. Draw the base circle and the follower in its lowest position.

2. Let P, the roller center, be the trace point. Calculate the length of the arc of travel of point P:

$$s_T = \frac{\pi}{180}\,\phi_T R \tag{6-4}$$

3. Construct the follower displacement diagram, using length s_T as the follower travel.

4. Draw radial lines dividing the cam into angular intervals corresponding to those used in the displacement diagram. Number in a direction opposite that of the cam's rotation.

5. Locate points 1′, 2′, 3′, etc. along the arc of follower movement by transferring corresponding follower rises from the displacement diagram to the arc with dividers, using small chordal lengths to minimize error.

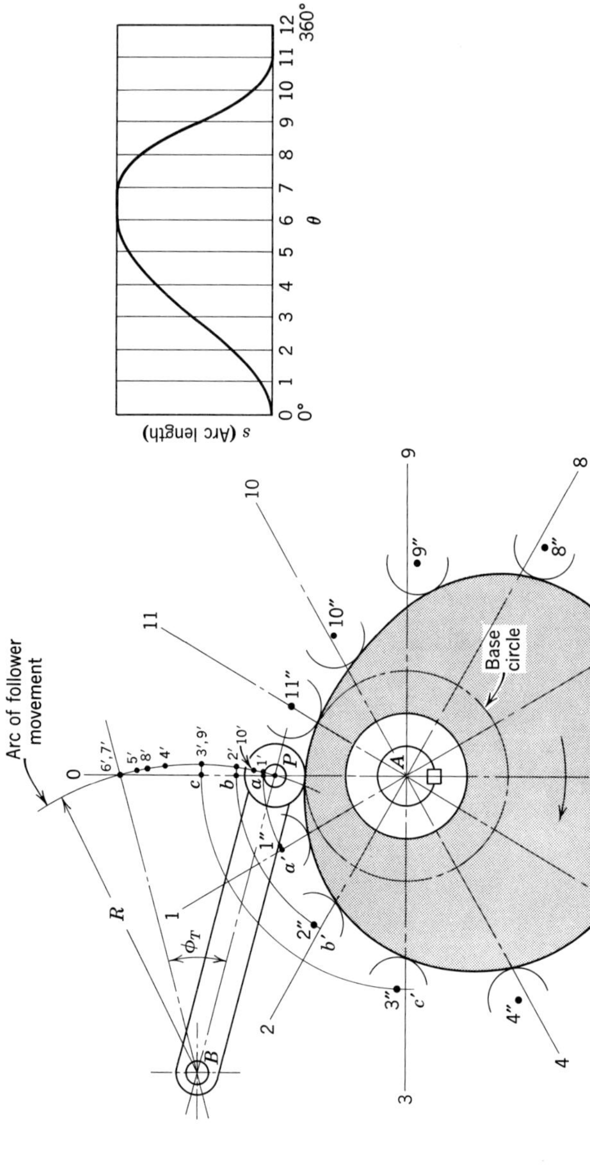

Fig. 6-20. Radial cam—oscillating roller follower.

6. With A as center and $A1'$ as radius, draw arc aa' from radial line 0 to radial line 1. With A as center and $A2'$ as radius, draw arc bb' from radial line 0 to radial line 2. Repeat for each point on the arc of follower travel.

7. Set dividers to offset $a1'$ and lay off from a' to locate 1″. Set $b2'$ from b' to locate 2″, $c3'$ from c' to locate 3″, and so on. Note that the directions of $a1'$ and $a'1''$, or of $b2'$ and $b'2''$, etc., must be consistent. Points 1″, 2″, 3″, etc. are the positions of the roller center, P, relative to the cam.

8. Draw the roller profiles and the cam profile.

6-14. Radial Cam—Oscillating Flat-Face Follower

Figure 6-21 gives an example of this construction. It is assumed that the base circle diameter, follower pivot location, follower dimensions, and follower angle of travel were either given or selected.

1. Draw the base circle and the follower in its lowest position.
2. Select a point P on the follower face to be the trace point. Calculate the length of its arc of travel, using Eq. 6-4.
3. Construct the follower displacement diagram, using length of s_T as follower travel.
4. Draw radial lines dividing the cam into angular intervals corresponding to those used in the displacement diagram.
5. Locate points 1′, 2′, 3′, etc. along the arc of follower movement by transferring corresponding follower rises from the displacement diagram to the arc with dividers, using small chordal lengths to minimize error.
6. At B_0, the follower pivot, draw a circle tangent to the follower face extended. This is called the *tangent circle*.
7. Draw radial lines dividing the cam into angular intervals corresponding to those used in the follower displacement diagram, *starting from* AB_0 and proceeding in a direction opposite that of the cam's rotation.
8. With A as center and AB_0 as radius, draw a circle intersecting these radial lines at B_1, B_2, B_3, etc., the alternate positions of the follower pivot relative to the cam.
9. At each of these positions draw the tangent circle.
10. Draw an arc of radius $A1'$ from A as center and of indefinite length. With B_1 as center and R as radius ($R = B_0P$), strike an arc crossing the first arc at 1″. Draw an arc of radius $A2'$ from A as center. From B_2 as center and with R as radius, strike an arc crossing the one just drawn at 2″, In the same manner locate 3″, 4″, etc., the positions of the trace point, P. relative to the cam.
11. To locate the relative positions of the follower face, draw a line from

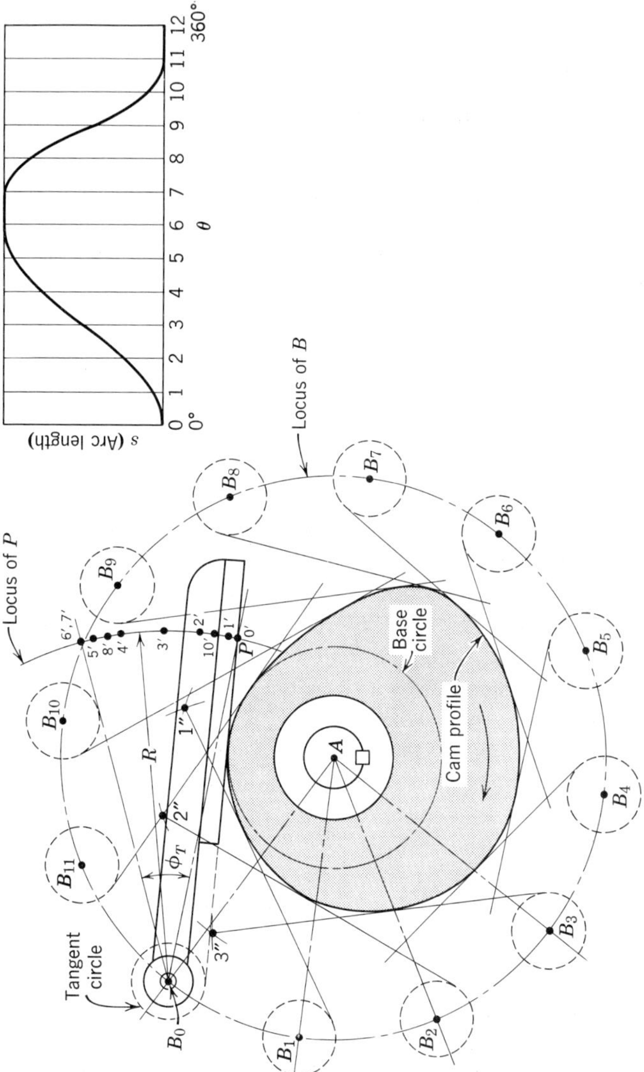

Fig. 6-21. Radial cam—oscillating, flat-face follower.

1″ and tangent to the circle at B_1, another from 2″ and tangent to the circle at B_2, and so on.

12. Draw the cam profile tangent to each of these follower profiles.

6-15. Cam Size

Generally it is desirable to make the cam small. A small cam not only implies compactness, but also less trouble with stress, noise, and vibration at the higher speeds. However, there are three factors which place limits on the minimum size of a cam: (1) pressure angle, (2) undercutting, and (3) hub size.

Pressure angle. The significance of the pressure angle is illustrated in Fig. 6-22. At the instant shown, the force exerted on the follower by the cam is along the common normal to the surfaces at the point of contact. And the angle between this force, F, and the center line of the follower is the pressure angle, α.

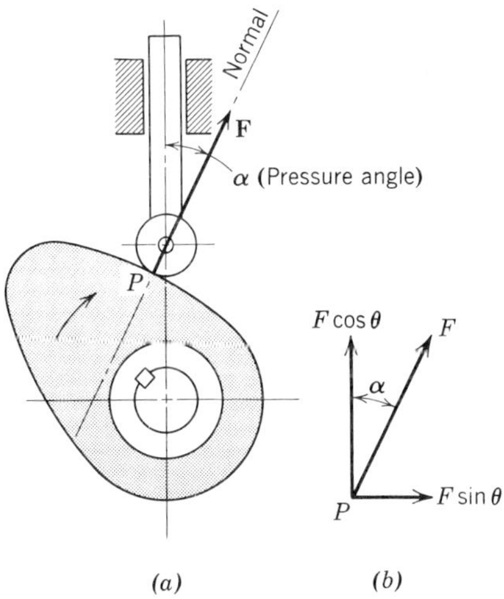

Fig. 6-22. Pressure angle.

The force, F, has been resolved, in Fig. 6-22b, into components along and normal to the path of travel of the follower. The component normal to the path of travel causes friction between the follower and its guide and, if large enough, causes binding. Since the magnitude of this component increases as the pressure angle increases, it is necessary to keep the pressure angle small. Designers usually try to keep the maximum pressure angle under 30°.

If a cam layout has been made and the pressure angle found to be excessive at some cam position, design changes must be made. A larger base circle may be used, the follower offset changed, or a different type of follower displacement curve employed. To minimize the effects of large pressure angles, the follower should be made as rigid as possible, the coefficient of friction between the follower and its guide be made as low as possible, and the follower overhang (distance the follower extends beyond the guide) kept as low as possible.

It is one of the advantages of the flat-face follower that pressure angle is not a critical problem. The pressure angle for a flat-face follower in which the face is normal to the follower center line is substantially zero. Only friction causes side thrust.

An oscillating follower also is less critical in design. Large pressure angles cause less difficulty than with translating roller followers, and often the effects of large pressure angles can be reduced simply by relocating the follower pivot.

Undercutting. Undercutting occurs when the cam curvature is such that it cannot guide the follower in its prescribed path. Examples of this problem are shown in Fig. 6-23. In Fig. 6-23*a* the radius of curvature of a

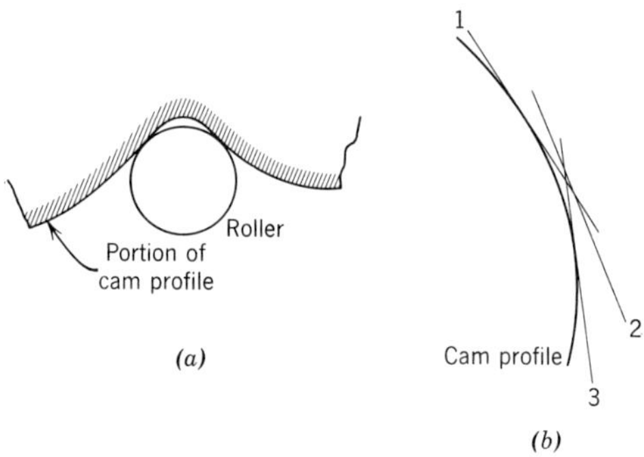

Fig. 6-23. Undercutting.

portion of the cam profile is less than the radius of the roller follower. Hence the roller cannot follow the cam profile. In Fig. 6-23*b* a portion of a cam profile and three positions of a flat-face follower are shown. When the cam profile is drawn tangent to the follower profiles at positions 1 and 3, it cannot touch the position 2 profile. The cam cannot, therefore, guide the follower in the desired manner.

Two common ways to revise a cam design to eliminate undercut are to reduce the roller diameter or to increase the base circle diameter. The easiest solution is usually to increase the cam size.

Hub size. Once the size of the cam shaft has been established, the hub size of the cam can be determined. This is generally done by a rule of thumb. For cast-iron cams, Rothbart (Ref. 2, p. 90) cites the following rule (Fig. 6-24):

$$D_H = 1.75D_S + 0.25 \tag{6-5}$$

where D_H = hub diameter, in.
D_S = shaft diameter, in.

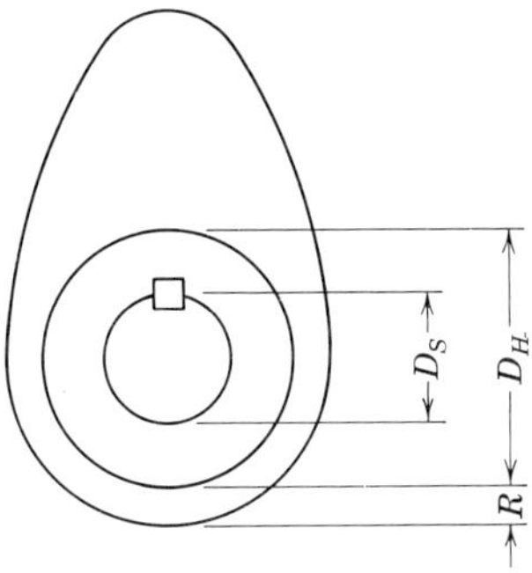

Fig. 6-24. Hub size.

For steel cams, the hub diameter may be made a little smaller. In either case the dimension R should be made at least $\frac{1}{8}$ in. The keyway should be located in the largest part of the cam.

The hub size and the dimension R place lower limits on the practical diameter of the base circle.

6-16. Cam Fabrication

Some cams can be cut on a milling machine in which the rotation of the cam blank on the spindle of the dividing head is coordinated with the feed of the machine table. For example, a uniform rise profile on a cam can be cut by a uniform rotation of the cam blank combined with a uniform feed of the table. A uniform rise profile on a radial cam is being cut in Fig. 6-25, and on a face cam in Fig. 6-27. In a similar manner, the slots on a drum cam are being cut in Fig. 6-28.

In most cases, however, the cam profile must be generated by incremental settings, that is, the contour is cut in steps, each step being a cut made at a particular cam position and cutter position. This leaves a scalloped profile, Fig. 6-29, which must then be finished by hand-filing or grinding the ridges left by the machining process. In Fig. 6-26 a pair of radial cams are being cut by incremental settings on a single machine set-up.

Fig. 6-25. Milling a uniform rise cam profile. (Courtesy The Cincinnati Milling Machine Co., Cincinnati, Ohio.)

Fig. 6-26. Milling radial cams by incremental settings. (Courtesy The Cincinnati Milling Machine Co., Cincinnati, Ohio.)

Fig. 6-27. Milling the uniform rise of a complex face cam. (Courtesy The Cincinnati Milling Machine Co., Cincinnati, Ohio.)

Fig. 6-28. Milling drum cam. (Courtesy The Cincinnati Milling Machine Co., Cincinnati, Ohio.)

The ridge height, e, on a radial cam (Fig. 6-29) varies (Ref. 3, p. 509):

1. directly as the square of the radial distance between the cam and cutter centers,
2. directly as the square of the angular cam displacement between adjacent cuts, and
3. inversely as the cutter radius.

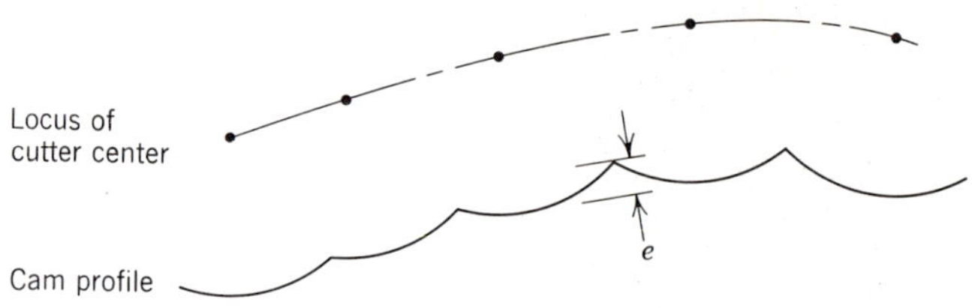

Fig. 6-29. Scalloped cam profile resulting from milling with incremental settings (greatly exaggerated).

The ridge height can thus be controlled by regulating the angular displacements between adjacent cuts or the cutter radius. For cam mechanisms with roller followers the cutter size is controlled by the roller size; thus the practical alternative is to regulate the cam displacements between cuts.

Since the ridge height, e, varies also as the square of the radial distance between the cam and cutter centers, larger cam displacements can be used at the smaller portions of the cam profile.

6-17. Profile Construction for Pure Rolling Contact

Two links oscillating about fixed centers and having pure rolling contact are shown in Fig. 6-30. For pure rolling contact to occur between these two links, three conditions must be met:

1. The lengths of the arcs of action on each body must be equal.
2. The point of contact must always be on the line of centers.
3. The sum of the distances from the centers of rotation to the point of contact must always equal the center distance.

In Fig. 6-30 the links are in contact at point P. Points c and c'' are points on the two profiles which will eventually come into contact. The first of the above three rules states that the arc lengths Pc and Pc'' must be the same. The second rule states that, when c and c'' come into contact, it must occur on the line of centers (O_2O_3). And the third rule implies that

$$O_2c + O_3c'' = O_2O_3$$

Suppose now that the profile of link 2 is given, and it is desired to construct the profile of link 3 so that pure rolling contact can occur between the two links. The following construction can be used:

1. With a divider lay out points equally spaced on the given profile, points a, b, c, etc.
2. Revolve each point about O_2 to the line of centers. Point a revolves to a', b to b', etc.

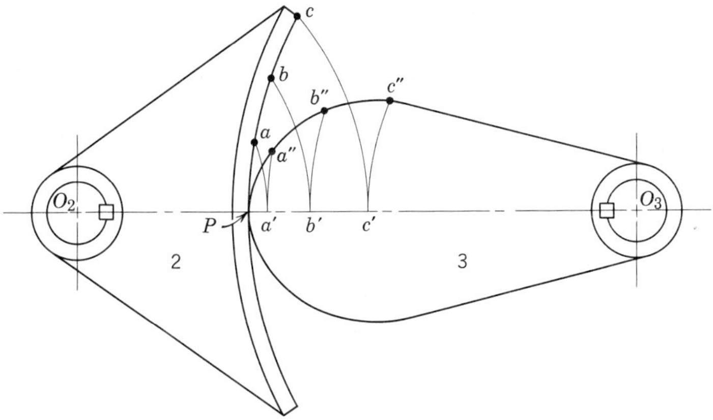

Fig. 6-30. Profile construction for rolling contact.

3. About O_3 strike arcs through points a', b', c', etc.
4. Set the dividers to the spacing of Step 1 and from P locate a'' on arc $a'a''$, from a'' locate b'' on arc $b'b''$, from b'' locate c'' on arc $c'c''$, and so on. Points a'', b'', c'', etc. determine the profile of link 3.

The above construction is only approximate, since it was assumed that chord lengths Pa and Pa'' are equal, that chord lengths ab and $a''b''$ are equal, and so on. However, if the original spacing is not too great, sufficient accuracy is possible.

A more complete treatment of the construction of profiles for pure rolling contact is given by Hinkle (Ref. 1).

REFERENCES

1. Rolland T. Hinkle, *Kinematics of Machines*, 2nd ed., Art. 8.7, Prentice-Hall, Englewood Cliffs, N.J., 1960.
2. Harold A. Rothbart, *Cams: Design, Dynamics, and Accuracy*, John Wiley and Sons, New York, 1956.
3. *A Treatise on Milling and Milling Machines*, 3rd ed., ch. 14, The Cincinnati Milling Machine Company, Cincinnati, Ohio, 1951.

PROBLEMS

In Probs. 6-1 to 6-10, follower displacement diagrams are to be plotted. The scales should be: follower displacement, full size; cam displacement, 60°/in. Divide the cam displacement axis into 15° intervals.

6-1. 0° to 180°, rise 2 in. with simple harmonic motion. 180° to 210°, dwell.
210° to 330°, fall 2 in. with simple harmonic motion.
330° to 360°, dwell.

6-2. 0° to 150°, rise 2 in. with constant acceleration.
150° to 330°, fall 2 in. with constant acceleration.
330° to 360°, dwell.

6-3. 0° to 150°, rise 1.5 in. with modified constant velocity (30°, 90°, 30°).
150° to 180°, dwell.
180° to 345°, fall 1.5 in. with modified constant velocity (45°, 75°, 45°).
345° to 360°, dwell.

6-4. 0° to 180°, rise 3 in. with cycloidal motion
180° to 210°, dwell.
210° to 330°, fall 3 in. with cycloidal motion.
330° to 360°, dwell.

6-5. 0° to 120°, rise 2.5 in. with simple harmonic motion.
120° to 330°, fall 2.5 in. with constant acceleration.
330° to 360°, dwell.

6-6. 0° to 120°, rise 2 in. with cycloidal motion.
120° to 150°, dwell.
150° to 300°, fall 2 in. with modified constant velocity (30°, 90°, 30°).
300° to 360°, dwell.

6-7. 0° to 180°, rise 2.5 in. with modified constant velocity (45°, 90°, 45°).
180° to 360°, fall with constant acceleration.

6-8. 0° to 240°, rise 2 in. with cycloidal motion.
240° to 300°, dwell.
300° to 360°, fall 2 in. with constant acceleration.

6-9. 0° to 90°, rise 1 in. with simple harmonic motion.
90° to 120°, dwell.
120° to 210°, rise 1 in. with simple harmonic motion.
210° to 240°, dwell.
240° to 330°, fall 2 in. with constant acceleration.
330° to 360°, dwell.

6-10. 0° to 120°, rise 2 in. with cycloidal motion.
120° to 150°, dwell.
150° to 240°, fall 1 in. with constant acceleration.
240° to 270°, dwell.
270° to 360°, fall 1 in. with simple harmonic motion.

In Probs. 6-11 to 6-16, cam profiles are to be constructed. In each case use one of the displacement diagrams of Probs. 6-1 to 6-10 as directed by the instructor.

6-11. In-line, knife-edge follower.
Base circle diameter, 2 in.
Cam turns clockwise.

6-12. In-line, roller follower.
Base circle diameter, $2\frac{1}{4}$ in.
Roller diameter, $\frac{5}{8}$ in.
Cam turns counterclockwise.
After laying out the cam, determine the maximum pressure angle.

6-13. In-line, flat-face follower.
Base circle diameter, 2.5 in.
Cam turns clockwise.
After laying out the cam, determine the minimum follower diameter.

6-14. Roller follower offset $\frac{7}{8}$ in. to the left.
Base circle diameter, $2\frac{1}{4}$ in.
Roller diameter, $\frac{5}{8}$ in.
Cam turns clockwise.
After laying out the cam, determine the maximum pressure angle.

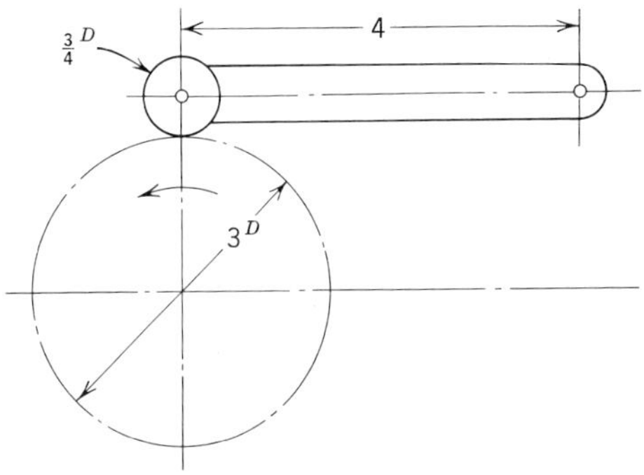

Fig. P6-15

6-15. Oscillating roller follower, Fig. P6-15.
Lay out the follower displacement along the arc of movement of the roller center.

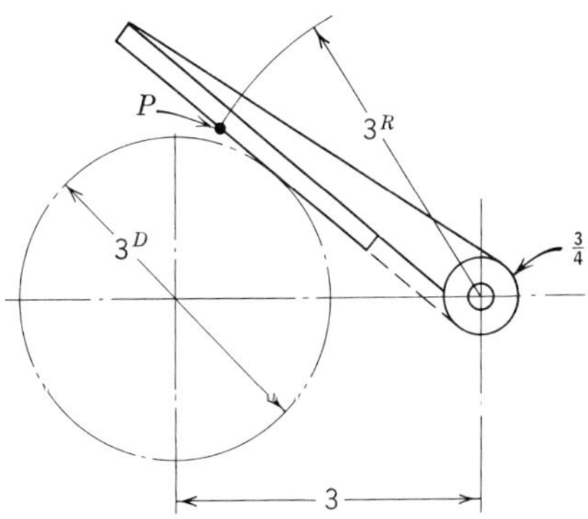

Fig. P6-16

6-16. Oscillating flat-face follower, Fig. P6-16.
Lay out the follower displacement along the arc of movement of point *P*.

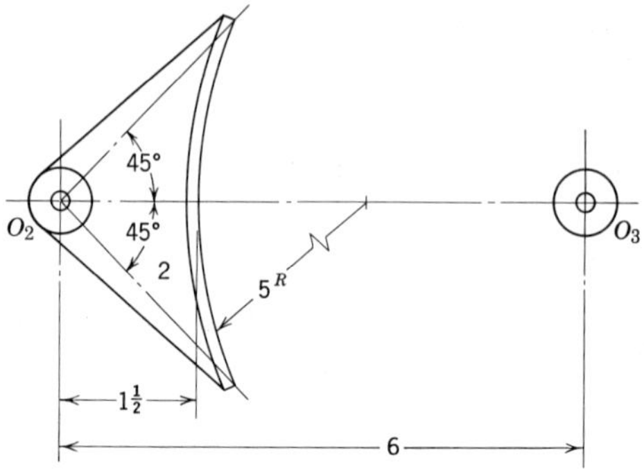

Fig. P6-17

6-17. Construct the profile of a link oscillating about pivot point O_3 which will have pure rolling contact with the link 2 shown in Fig. P6-17.

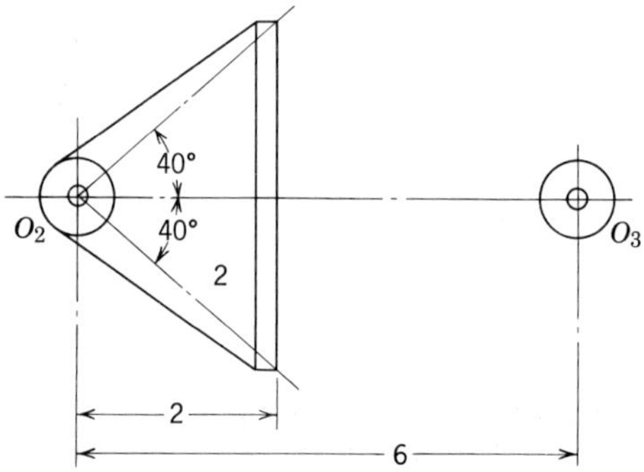

Fig. P6-18

6-18. Construct the profile of a link oscillating about pivot point O_3 which will have pure rolling contact with the link 2 shown in Fig. P6-18.

7 Gears

7-1. Introduction

Gears transmit motion between one shaft and another, or between a shaft and a slide, by means of successively engaging teeth.

Although gears had been used as early as the fourth century, B.C., it was not until the invention of methods for milling and for generating tooth profiles in the latter half of the nineteenth century that gears could be fabricated with sufficient accuracy to permit their use at high speeds and under heavy loads.

In this chapter primary emphasis will be given spur gears; the student will become acquainted with the fundamentals of gear theory through their study. What he learns about the theory of spur gears will prepare him for subsequent, more detailed studies of other types of gearing.

7-2. Classification of Gears

Gears can be conveniently classified by the relative positions of their shaft axes. Table 7-1 shows the types of gearing commonly used to connect shafts having each of the three possible shaft relationships: parallel, intersecting, or skew.

TABLE 7-1

Shaft Relationship	Type of Gearing
Parallel	Spur Helical Double helical Herringbone
Intersecting	Bevel
Skew	Helical Worm Hypoid

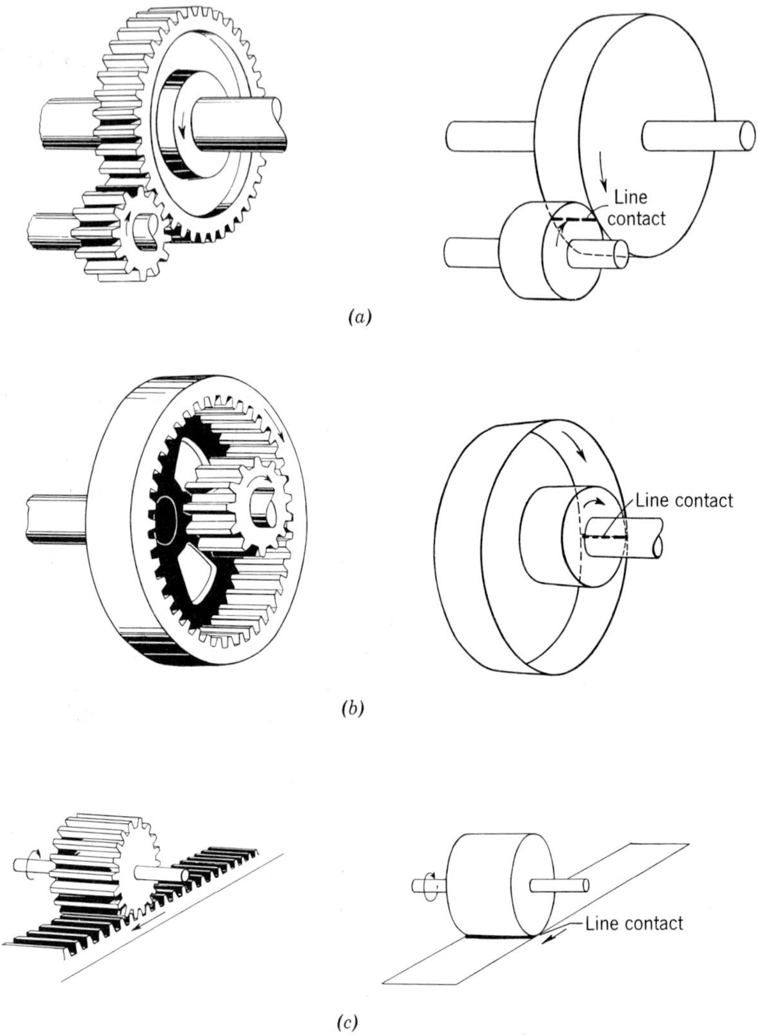

Fig. 7-1. Spur gears. (*a*) External spur gear and pinion. (*b*) Internal spur gear and pinion. (*c*) Spur pinion and rack. (Courtesy Socony-Mobil Oil Co., New York.)

Spur gears are illustrated in Fig. 7-1. Kinematically they are equivalent to a pair of rolling cylinders. The teeth are straight and parallel to the gear axis. If the gears have external contact, Fig. 7-1*a*, they turn in opposite directions. With internal contact, Fig. 7-1*b*, they turn in the same direction. The spur pinion and rack combination in Fig. 7-1*c* makes it possible to convert rotational to translatory motion, or vice versa. It is sometimes convenient to regard the rack as a spur gear with an infinitely large diameter.

Helical gears, Fig. 7-2, also can be used to connect parallel shafts. They, like spur gears, are equivalent to rolling cylinders. But their teeth are curved, each being helical in shape. This curving of the teeth results in smoother contact between mating teeth. Helical gears can be used at

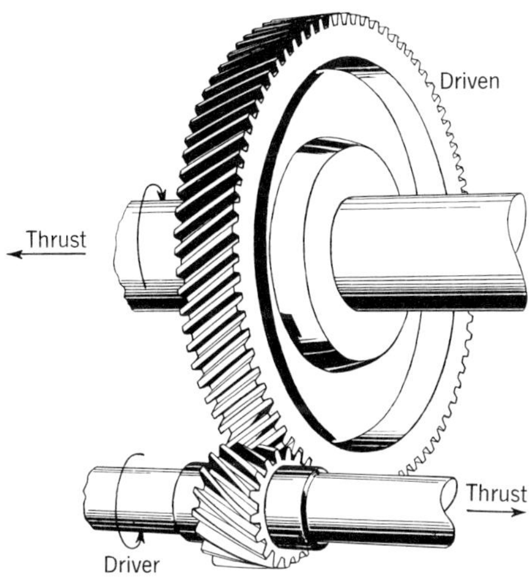

Fig. 7-2. Single-helical gear and pinion. (Courtesy Socony-Mobil Oil Co., New York.)

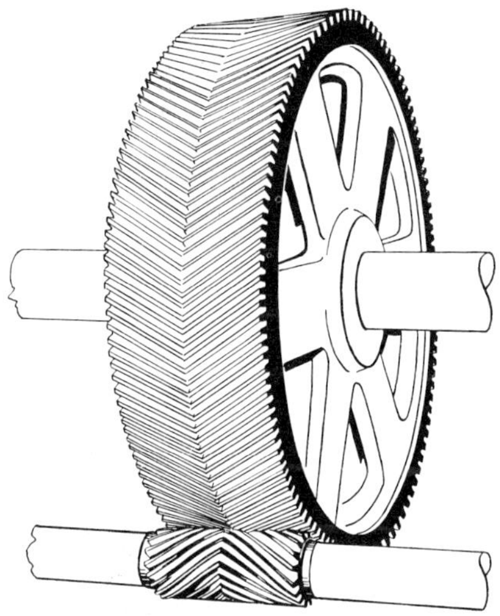

Fig. 7-3. Herringbone gear and pinion. (Courtesy Socony-Mobil Oil Co., New York.)

higher velocities than spur gears because of their ability to run more quietly. And they have, in addition, greater load-carrying capacity than spur gears of comparative size. A disadvantage arising from the curving of the teeth is the end thrust: the force between mating teeth has a component along the gear axis. The bearings and assemblies used in mounting helical gears must be capable of withstanding thrust loads.

A *continuous tooth herringbone gear*, Fig. 7-3, is equivalent to a pair of helical gears with opposite helix angles mounted side by side. This type of gear is superior to the single helical gear in that the axial thrusts of the two rows of teeth cancel each other out. They can be used at high speeds with a minimum of noise and vibration.

Double helical gears, like herringbone gears, are each equivalent to a pair of helical gears mounted side by side. Unlike the herringbone gears, the teeth of the two rows are not continuous; they are separated by a

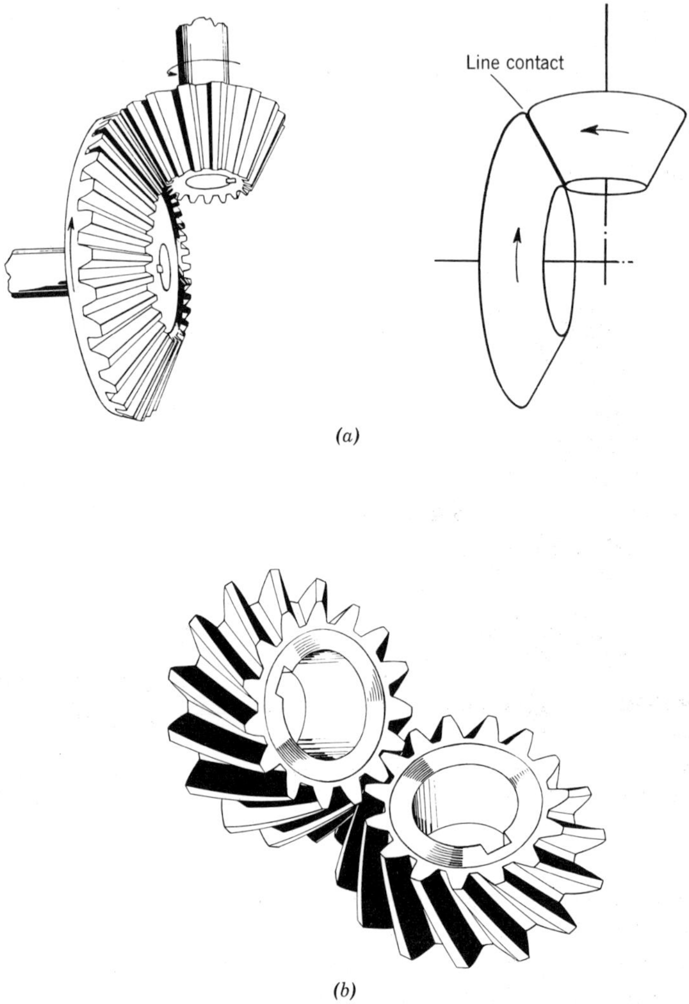

Fig. 7-4. Bevel gears. (*a*) Straight tooth bevel gears. (*b*) spiral bevel gears. (Courtesy Socony-Mobil Oil Co., New York.)

groove cut into the gear blank and used for tool runout. Double helical gears are used for ultra high speeds where high helix angles and tooth shaving are necessary.

Bevel gears, Fig. 7-4, are used to connect intersecting shafts. Kinematically they are equivalent to rolling cones. *Straight tooth bevel gears*, Fig. 7-4*a*, are generally used to connect shafts at right angles and running at low speeds (pitch line velocities under 1000 fpm). When the shafts connected are at right angles and running at the same speed, the gears are of the same size and are called *miter gears.*

Zerol bevel gears have teeth that are curved, but with a 0° spiral angle. Because the teeth are curved, they are quieter in action than straight tooth gears. And since they produce no more axial thrust than the straight tooth bevels, they can replace straight tooth bevel gears with no modifications in the bearings or supporting assemblies.

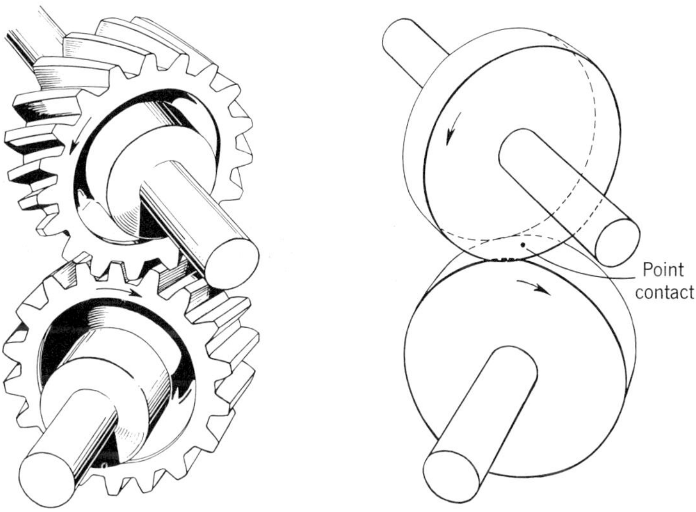

Fig. 7-5. Helical gears connecting skew shafts. (Courtesy Socony-Mobil Oil Co., New York.)

Spiral bevel gears, Fig. 7-4*b*, are smoother in action and more quiet than straight tooth bevel gears. They can be used at pitch-line velocities up to 8000 fpm. A spiral bevel pinion can have fewer teeth than a straight tooth bevel pinion of comparable size. But, like the helical gear, there is greater axial thrust, dictating more substantial bearings and supporting assemblies.

Helical gears, Fig. 7-5, used to connect skew shafts are sometimes called *spiral gears.* Although limited to light loads, they have a number of advantages. They can transmit motion between skew shafts at any angle. From a single direction of rotation of the driving shaft, the driven shaft can be made to turn in either direction merely by a suitable choice of the helix angles for the mating gears. Velocity ratios for an existing pair of skew shafts can be altered, without change in center distance, by substituting different gears.

Worm gear drives, Fig. 7-6, permit very high speed reduction ratios to be

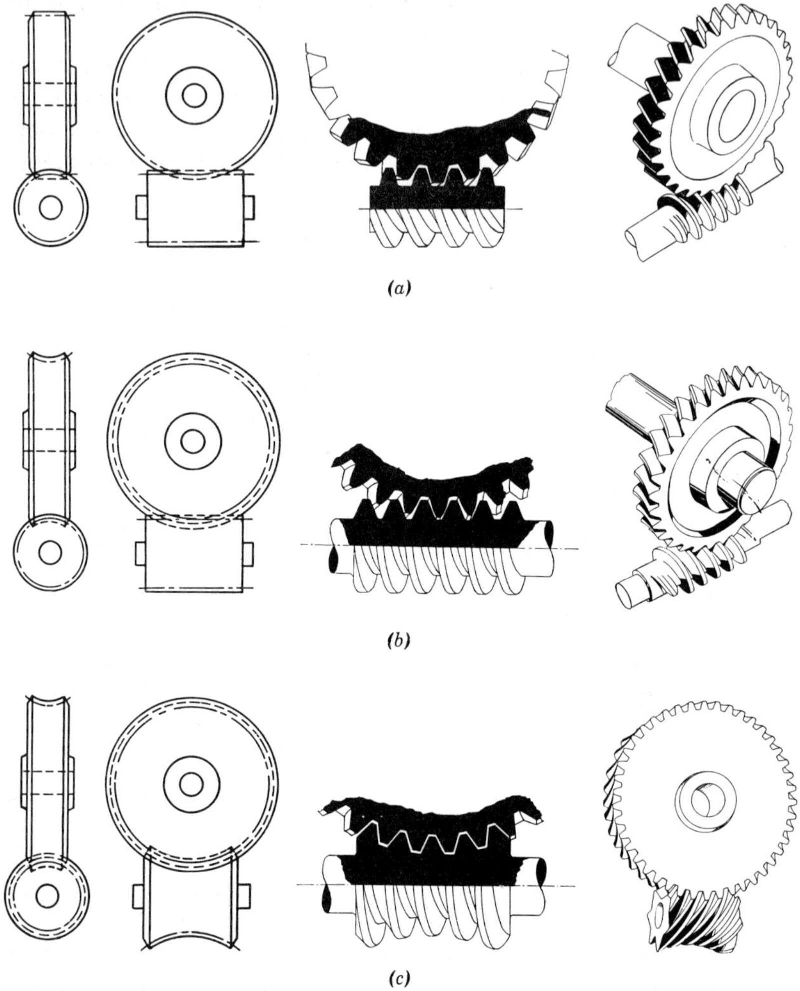

Fig. 7-6. Worm gear drives. (*a*) Nonthroated worm and gear. (*b*) Single-throated worm and gear. (*c*) Double-throated worm and gear. (Courtesy Socony-Mobil Oil Co., New York.)

achieved in a small center distance. They are used to connect skew shafts which are perpendicular. This type of drive can transmit heavy loads without shock or vibration, producing a smooth output free of pulsation. The chief disadvantage of a worm gear drive is its low mechanical efficiency. A large proportion of the input power is lost in friction. The capacity of a worm gear drive unit is limited by its ability to dissipate heat rather than by the strengths of its components.

For the nonthroated worm and gear design of Fig. 7-6*a*, the gear teeth are not throated and do not envelop the worm. Contact between the teeth is concentrated at a point. In the single-throated design of Fig. 7-6*b*, the gear teeth are curved to envelop the worm. This results in line contact between mating teeth. And the double-throated design of Fig.

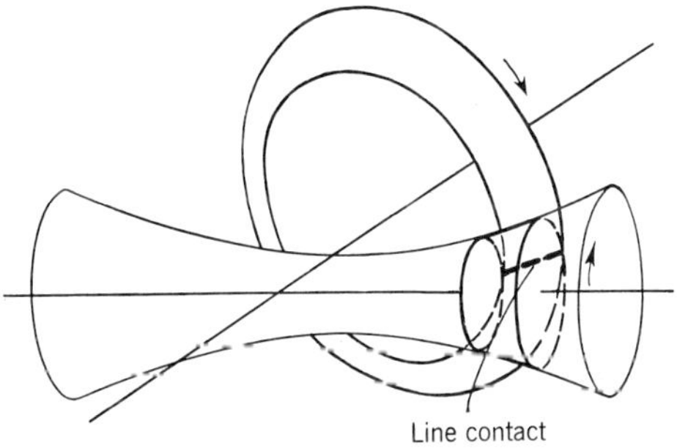

Fig. 7-7. Hypoid gear and pinion. (Courtesy Socony-Mobil Oil Co., New York.)

7-6*c* results in area contact. Both the worm and the gear have been made with concave outer surfaces to envelop each other.

Hypoid gears, Fig. 7-7, look very much like spiral bevel gears, but they connect skew shafts. For a given speed ratio the hypoid pinion will be larger and stronger than a spiral bevel pinion. And the hypoid pair will run more smoothly and quietly than the spiral bevel pairs. The shafts of the hypoid gears can be made to pass each other, permitting bearings to be located at both sides of each gear. An additional advantage of this fact is that a single drive shaft can drive a series of machines or parts of a machine from a single source of power. This has been made use of in steel mills, printing machines, and paper mills.

7-3. Law of Gearing

The law of gearing will be demonstrated by reference to Fig. 7-8. Here two links are in direct contact. The instant center method of velocity analysis can be applied to determine the ratio of the angular speeds of the two links.

It is known that when two bodies are in direct contact, their common instant center lies at the intersection of the common normal at their point

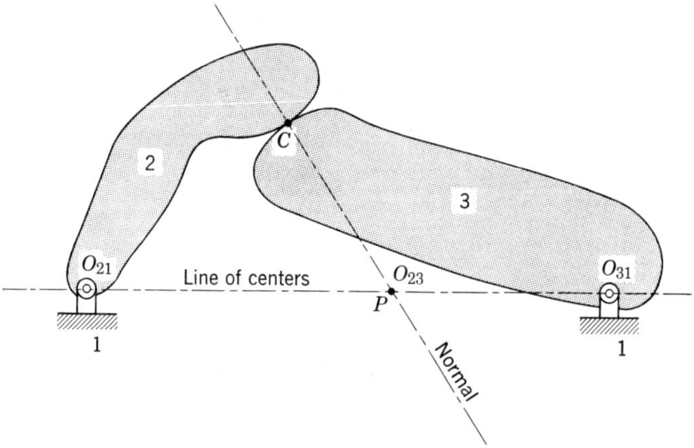

Fig. 7-8. Law of gearing.

of contact and their line of centers. In Fig. 7-8 the common instant center, O_{23}, has been found where the common normal at point C intersects the line of centers, $O_{21}O_{31}$. In gearing this point, O_{23}, is called the *pitch point*. It has also been labeled "P" on the drawing.

The ratio of the speeds of the two links is

$$\frac{\omega_3}{\omega_2} = \frac{O_{21}O_{23}}{O_{31}O_{23}}$$

The usual requirement in gearing is that this ratio be fixed. But if the centers of rotation of the two links are fixed points, then both the sum and the ratio of distances $O_{21}P$ and $O_{31}P$ are constants. And this in turn implies that point P is fixed.

If the mating profiles on links 2 and 3 in Fig. 7-8 are regarded as the profiles of mating teeth, the following law can be stated: *The common normal to the tooth profiles at their point of contact must always pass through the pitch point.*

Tooth profiles which satisfy this criterion are said to be *conjugate*. Cycloidal and involute curves have been used for gear tooth profiles, though the involute profile is nearly always used now.

7-4. Spur Gear Terminology

The terminology of spur gears is illustrated in Figs. 7-9 and 7-10. The definitions of many of the terms commonly used in discussions of spur gears are given below. Additional terms will be introduced in later articles as needed.

The *pitch circle* is the theoretical circle upon which gear calculations are based. The pitch circles of a pair of mating spur gears are the kinematically

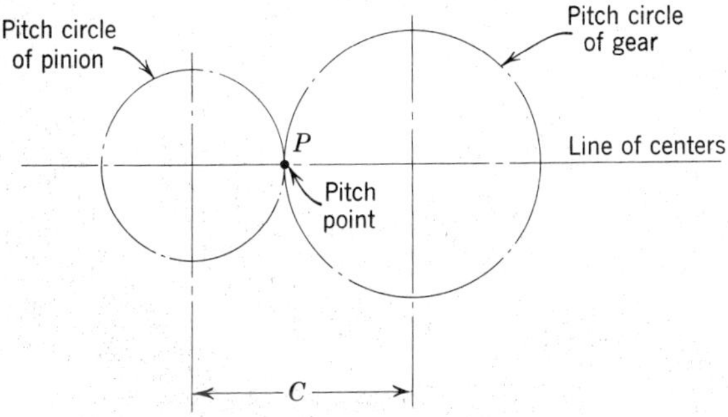

Fig. 7-9. Spur gear terminology.

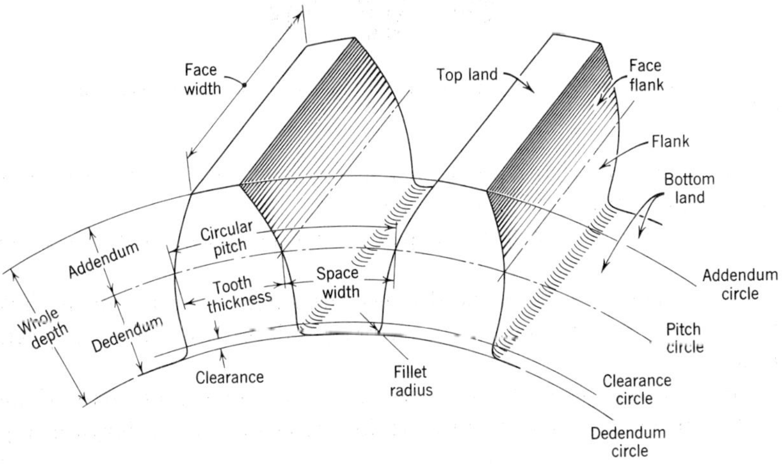

Fig. 7-10. Spur gear terminology.

equivalent rolling circles. The *pitch diameter*, D, of a gear is the diameter of the pitch circle.

The *line of centers* is the line through the centers of rotation of a pair of mating gears.

Fig. 7-11. Tooth sizes for standard diametral pitches. (Courtesy Barber-Colman Co., Rockford, Ill.)

The *pitch point* is the point of tangency between the pitch circles of mating gears. It is on the line of centers.

The *pinion* is the smaller of a pair of mating gears and the *gear* the larger.

The *circular pitch*, p, is the distance measured along the pitch circle from a point on one tooth to the corresponding point on the adjacent tooth.

The *diametral pitch*, P, is the ratio of the number of teeth to the pitch diameter. A comparison of various sizes of teeth is shown in Fig. 7-11.

From the definitions of circular pitch and diametral pitch given above, the following two equations can be written:

$$p = \frac{\pi D}{N} \tag{7-1}$$

$$P = \frac{N}{D} \tag{7-2}$$

where D = pitch diameter, in.
N = number of teeth
p = circular pitch, in.
P = diametral pitch, in.$^{-1}$

Equation 7-1 can be rewritten as

$$p = \frac{\pi}{N/D}$$

which, after substituting from Eq. 7-2, becomes

$$p = \frac{\pi}{P}$$

or

$$pP = \pi \tag{7-3}$$

Equation 7-3 is very useful. If either the circular pitch or the diametral pitch is known for a gear, the other can be found with no further knowledge of the gear's dimensions.

The *metric module* is the ratio of the pitch diameter in millimeters to the number of teeth. The *English module* is the ratio of the pitch diameter in inches to the number of teeth. English module and diametral pitch are reciprocal quantities.

The *gear ratio*, m_g, is the ratio of the number of teeth on the gear to that on the pinion. That is,

$$m_g = \frac{N_g}{N_p} \tag{7-4}$$

The *addendum*, a, is the radial distance between the pitch circle and the top of the tooth

The *dedendum*, b, is the radial distance between the pitch circle and the bottom of the tooth space.

The *clearance* is the amount by which the dedendum of one gear exceeds the addendum of the mating gear.

The *whole depth*, h_t, is the total depth of the tooth space. It is the sum of the addendum and the dedendum or of the working depth and the clearance.

The *working depth*, h_w, is the maximum depth to which a tooth penetrates the tooth space of the mating gear. It is the sum of the addendums of the two gears.

The *circular tooth thickness*, t, is the tooth thickness measured along the pitch circle.

The *space width* is the width of the tooth space measured along the pitch circle.

The *backlash, B*, is the amount by which the space width exceeds the circular tooth thickness.

The *top land* is the surface of the top of the tooth.

The *bottom land* is the surface of the gear between the fillets of adjacent teeth.

The *face* is that surface of the tooth which is between the top land and the pitch circle.

The *flank* is that surface of the tooth which is between the pitch circle and the bottom land. The flank includes the fillet.

The *fillet* is the concave portion of the tooth profile where the flank joins the bottom land. The fillet in generated gears is not circular, but the approximate radius of curvature is the *fillet radius*, r_f.

The *face width*, f, is the length of the tooth measured parallel to the gear axis.

7-5. Velocity Ratio

From an inspection of Fig. 7-9 it is clear that

$$C = \frac{D_p}{2} + \frac{D_g}{2}$$

or

$$2C = D_p + D_g \tag{7-5}$$

From Eq. 7-2,

$$D = \frac{N}{P}$$

Substituting for D in Eq. 7-5,

$$2C = \frac{N_p}{P} + \frac{N_g}{P}$$

or

$$2PC = N_p + N_g \tag{7-6}$$

Kinematically spur gears are equivalent to rolling cylinders, and the speeds of rolling cylinders vary inversely as their diameters. Therefore,

$$\frac{\omega_p}{\omega_g} = \frac{D_g}{D_p} \tag{7-7}$$

Substituting for D from Eq. 7-2,

$$\frac{\omega_p}{\omega_g} = \frac{N_g/P}{N_p/P}$$

or

$$\frac{\omega_p}{\omega_g} = \frac{N_g}{N_p} = m_g \tag{7-8}$$

Equation 7-6 can be rewritten as

$$2PC = N_p\left(1 + \frac{N_g}{N_p}\right)$$

Substituting for N_g/N_p from Eq. 7-4,

$$2PC = N_p(1 + m_g) \tag{7-9}$$

Equation 7-9 is useful in trial-and-error designs of gears. In Eqs. 7-5 to 7-9,

D_p = pitch diameter of pinion, in.
D_g = pitch diameter of gear, in.
C = center distance, in.
N_p = number of teeth on pinion
N_g = number of teeth on gear
P = diametral pitch, in.$^{-1}$
ω_p = speed of pinion, rad/sec or rpm
ω_g = speed of gear, rad/sec or rpm
m_g = gear ratio

Example 7-1. Spur gears are to connect two shafts 8 in. apart and turning at speeds of 550 rpm and 250 rpm. The teeth are to be of 6 diametral pitch. Determine: (1) the pitch diameters, (2) the numbers of teeth, and (3) the circular pitch.

Solution

1. From Eq. 7-5,

$$2(8) = D_p + D_g$$
$$16 = D_p + D_g$$

From Eq. 7-8,

$$\frac{550}{250} = \frac{D_g}{D_p}$$

Solving simultaneously for D_p and D_g

$$D_p = 5 \text{ in.}$$
$$D_g = 11 \text{ in.}$$

2. From Eq. 7-2,

$$N = PD$$
$$N_p = (6)(5)$$
$$= 30 \text{ teeth}$$
$$N_g = (6)(11)$$
$$= 66 \text{ teeth}$$

3. From Eq. 7-3,

$$p = \frac{\pi}{P}$$

$$= \frac{\pi}{6}$$

$$p = 0.5236 \text{ in.}$$

Example 7-2. Spur gears are to connect shafts 9 in. apart. The pinion turns at 1800 rpm and the gear speed is to be as close to 797 rpm as possible. The pinion shall have no fewer than 18 teeth, and the diametral pitch shall be one of the following: 3, 4, 5, 6. Find the most satisfactory solution.

Solution

1. By Eq. 7-9,

$$m_g = \frac{\omega_p}{\omega_g}$$

$$= \frac{1800}{797}$$

$$= 2.258$$

Substituting for m_g in Eq. 7-7,

$$2P(9) = N_p(1 + 2.258)$$

or

$$N_p = 5.524P$$

2. From Eq. 7-6,

$$N_p + N_g = 2P(9)$$

or

$$N_g = 18P - N_p$$

3. The above two equations for N_p and N_g contain one too many unknowns. Therefore a trial-and-error solution will be made. The criteria are: (*a*) the values of N_p and N_g, the numbers of teeth, must be whole numbers, (*b*) the number N_p must be at least 18, and (*c*) the value of the gear ratio, m_g, must be as close to 2.258 as possible. The method will be to try a value of P, calculate the corresponding value of N_p, rounding it off to a whole number, calculate the value of N_g, and, finally, recalculate the value of the gear ratio, m_g. The various solutions thus obtained will then be examined to determine which best fits the criteria.

4. The final calculations are shown in Table 7-2.

5. An examination of the results in Table 7-2 reveals that a diametral pitch of 3 results in a pinion which is too small, having only 17 teeth. Of the remaining solutions, $P = 4$ or $P = 6$ are each satisfactory. Each has a

TABLE 7-2

P	$N_p = 5.524P$	N_p (Rounded)	$18P$	$N_g = 18P - N_p$	$m_g = N_g/N_p$
3	16.57	17	54	37	2.176
4	22.1	22	72	50	2.273
5	27.6	28	90	62	2.215
6	33.1	33	108	75	2.273

proper pinion size, and each is equally close to the required gear ratio. For $P = 5$ the pinion size is satisfactory, but the gear ratio is not so close to the required value as the other solutions.

7-6. Involute Tooth Profile

The *involute* of a circle is the locus of the end of a string unwound from the circle and kept taut. An involute is shown in Fig. 7-12. The end of the string, T, was originally at T_0. At any position, say T_2, the string is tangent to the circle. The string length, T_2t_2, is necessarily the same as the arc length, T_0t_2. The circle from which the string is unwound is called the *base circle*.

Figure 7-13 shows two involutes that are tangent to each other, along

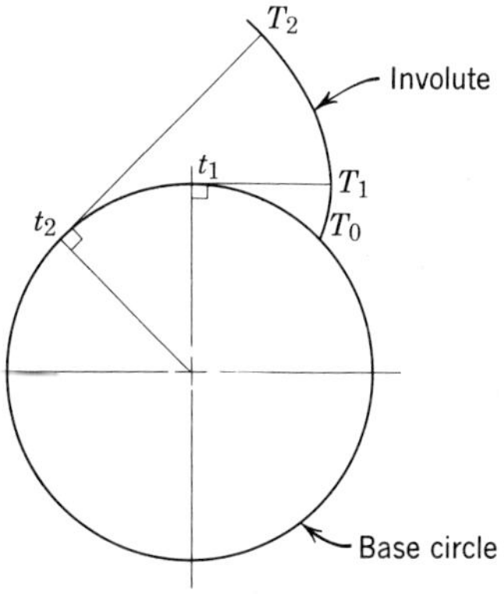

Fig. 7-12. The involute.

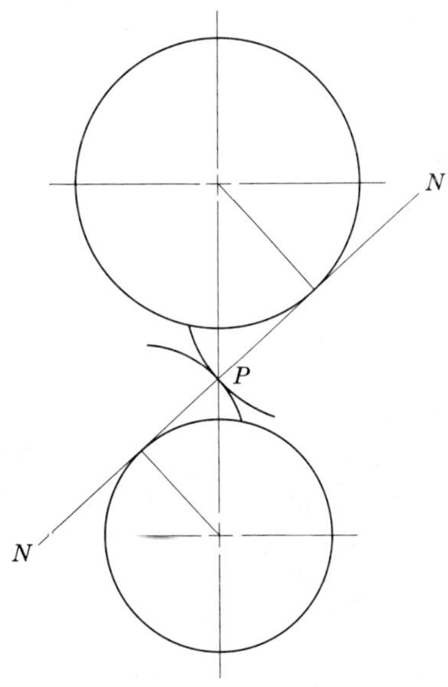

Fig. 7-13. Mating involutes.

with their base circles and their common normal. When two involutes are tangent, their common normal is tangent to both base circles. If the base circles are allowed to rotate about fixed centers of rotation, keeping the involutes tangent, the point of contact remains on the common normal, *N-N*, which remains fixed in position. Since the common normal remains fixed, the point where it crosses the line of centers is also fixed. Thus the involute tooth profile satisfies the basic law of gearing.

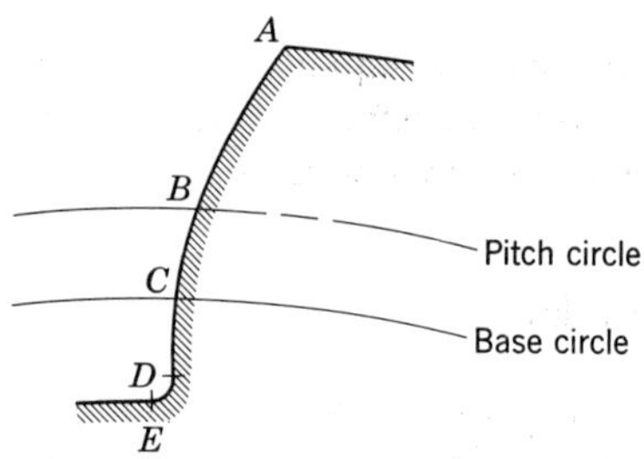

Fig. 7-14. Involute tooth profile.

A typical involute tooth profile appears in Fig. 7-14. The tooth profile extends from *A* to *E*, of which *AB* is the face and *BE* the flank. From the addendum circle, at *A*, the tooth is involute in shape down to the base circle at *C*. Thus the full face and a portion of the flank are involute in shape. Of the remaining tooth profile, *CD* is, nominally, a radial line and *DE* the fillet.

7-7. Constructing an Involute

The construction is illustrated in Fig. 7-15. It is assumed that the base circle and the initial point, 0, of the involute are given.

1. Lay off points equally spaced on the base circle. In the example of Fig. 7-15, these points are 0, 1, 2, 3, 4.
2. At each of these points draw a line tangent to the base circle.
3. On each tangent lay off a length equal to that of the corresponding arc. In the example, line length 1-1′ is made the same as arc length 1-0, line 2-2′ is made the same length as arc 2-0, etc.
4. Points 0, 1′, 2′, etc. are points on the involute. They are connected by use of a french curve.

In the above construction it was assumed that the initial point of the involute was known. Frequently some other point on the involute is known, along with the base circle. In Fig. 7-16, the base circle and point *A* are given. It is required to draw an involute passing through point *A*.

1. From *A* draw a line tangent to the base circle, locating carefully the tangency point, *a*.

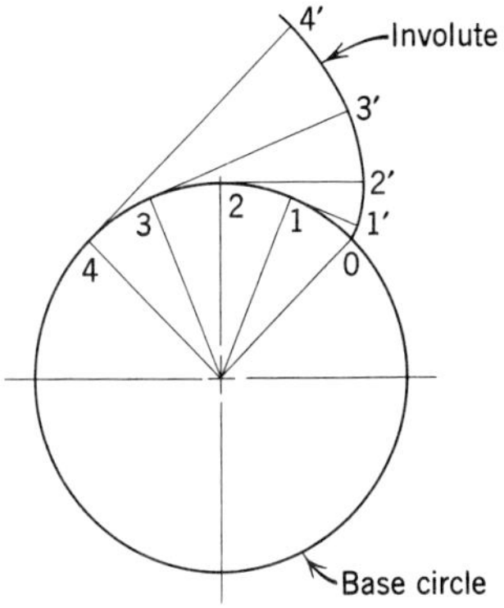

Fig. 7-15. Construction of the involute.

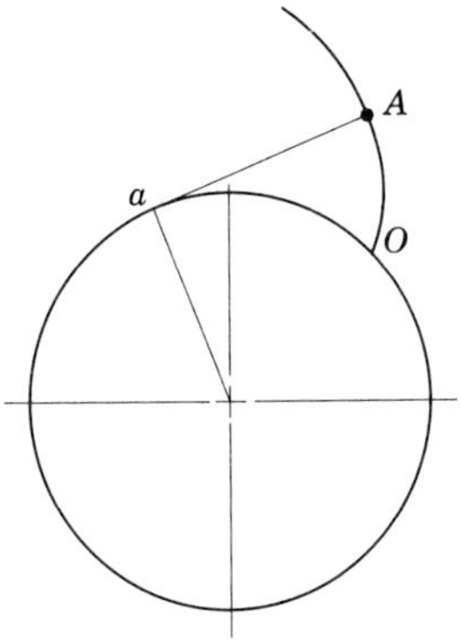

Fig. 7-16. Location of initial point on involute.

2. Find the initial point, *O*, of the involute on the base circle by making arc *aO* equal in length to line *a-A*. Set the dividers to a small chord length to avoid error.
3. Use the procedure given previously for constructing the involute.

7-8. Action of Spur Gear Teeth

In Fig. 7-17 the pitch and base circles have been laid out for a pair of mating spur gears. The common normal, *N-N*, to the tooth profiles at point of contact has been drawn tangent to both base circles. This line in gearing is called the *line of action*. The point of contact between mating teeth is always on this line.

The angle between the line of action, *N-N*, and the common tangent, *T-T*, to the pitch circles is called the *pressure angle*. This name derives from the fact that the force vector between mating teeth lies along this line.

The line of action is tangent to the driver's base circle at *A*. Thus O_2AP is a right triangle. The pitch radius of the driver, *R*, is O_2P, and its base circle radius, R_b, is O_2A. Thus,

$$R_b = R \cos \phi \tag{7-10}$$

Equation 7-10 permits the calculation of the base circle radius if the pitch circle size and pressure angle are known.

In American industrial practice, two pressure angles for spur gears have been standardized, 14°30′ and 20°. Other pressure angles are sometimes used where the special cutters are available.

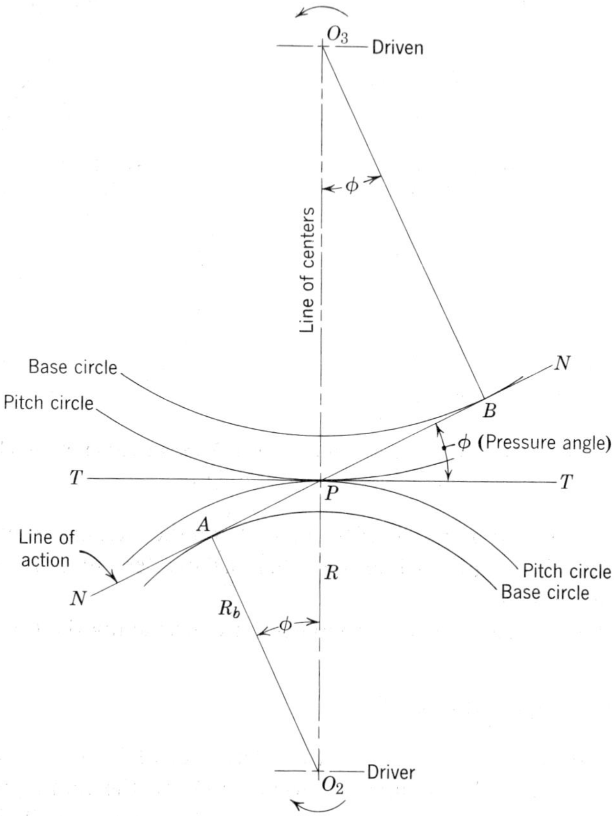

Fig. 7-17. Pressure angle and base circles.

To locate accurately the tangency points, A and B, between the line of action and the base circles:

1. Draw lines O_2A and O_3B (Fig. 7-17) each at angle ϕ from the line of centers.
2. Draw the line of action making an angle ϕ with the common tangent, T-T.
3. Where the line of action crosses line O_2A is A, and where it crosses O_3B is B.

Once points A and B have been located, the base circles can be drawn in. This makes it unnecessary to calculate their radii.

Points of initial and final contact. The line of centers, line of action, and addendum circles for a pair of mating spur gears have been drawn in Fig. 7-18. The point of contact between involute teeth is always on the line of

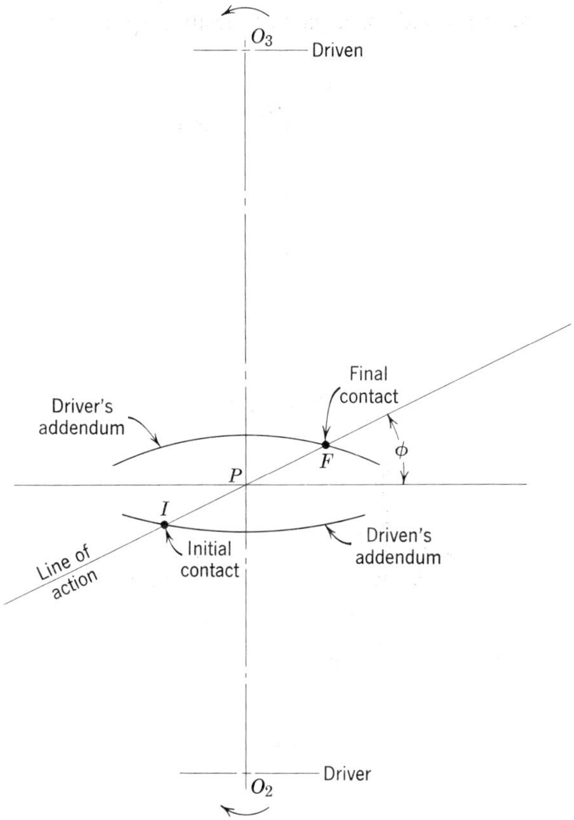

Fig. 7-18. Points of initial and final contact.

action. The initial point of contact occurs when the tip of the driven tooth reaches the line of action. The final point of contact occurs when the tip of the driver tooth reaches the line of action. The following rules can be used in finding these points:

1. The point of initial contact is at the intersection of the driven gear's addendum circle and the line of action.
2. The point of final contact occurs at the intersection of the driver's addendum circle with the line of action.

In the example, Fig. 7-18, the point of initial contact is I, and the point of final contact is F. Each pair of teeth comes into contact at I, remains in contact until F, then separates. Line IF is the *path of action*. The length of line IF is Z.

Angles and arcs of action. In Fig. 7-19 the points of initial and final contact have been located for a pair of meshing spur gears. In addition,

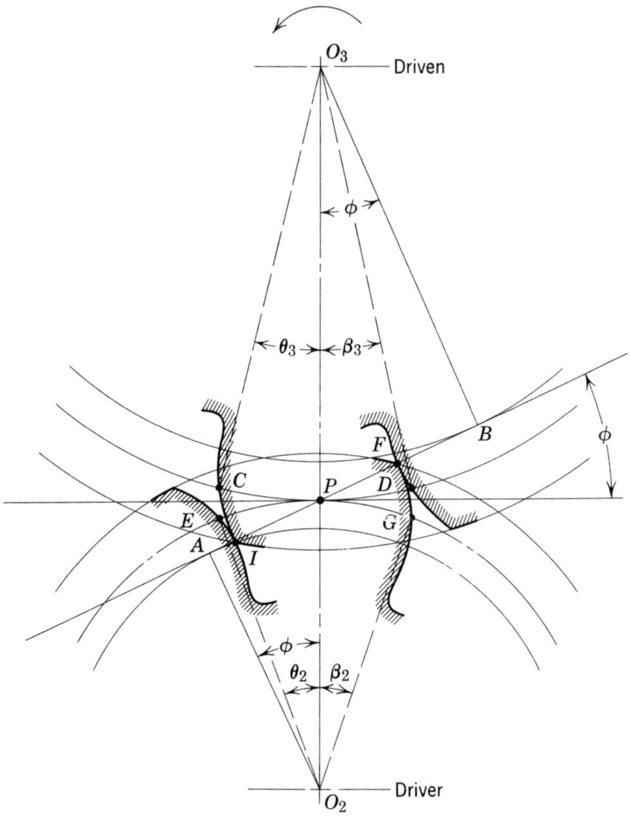

Fig. 7-19. Tooth action.

the profiles of the mating teeth have been drawn through point I, the point of initial contact. The same pair of teeth have been redrawn in their position of final contact, at F.

The *arc of action* for a gear is the arc of pitch circle through which a tooth travels from its position of initial contact to its position of final contact. For the driver in Fig. 7-19, the arc of action is EPG, and the driven gear's arc of action is CPD.

The *arc of approach* is the arc of pitch circle through which a tooth travels from its position of initial contact to its position at the pitch point. *EP* is the arc of approach for the driver, and *CP* for the driven, in Fig. 7-19.

The *arc of recess* is the arc of pitch circle through which a tooth travels from its position at the pitch point to its position of final contact. *PG* is the arc of recess for the driver, and *PD* for the driven, in Fig. 7-19.

The arc of action is the sum of the arcs of approach and recess. The arc of action is the same for each of a pair of mating gears.

The central angles subtended by the arc of action, arc of approach, and arc of recess for any gear are, respectively, the *angle of action*, *angle of approach*, and *angle of recess* for that gear. The angle of action is the sum of the angles of approach and recess. The angles of action are not, in general, the same for each of a pair of mating spur gears.

Contact ratio. The contact ratio, m_p, is a measure of the number of pairs of teeth in contact. The contact ratio can be found in two ways, either as the ratio of the arc of action to the circular pitch or as the ratio of the length of the path of action to the base pitch. That is,

$$m_p = \frac{Q}{p} \tag{7-11}$$

or

$$m_p = \frac{Z}{p_b} \tag{7-12}$$

where

$$p_b = p \cos \phi \tag{7-13}$$

In Eqs. 7-11 to 7-13,

m_p = contact ratio
Q = length of arc of action, in.
p = circular pitch, in.
Z = length of path of action, in.
p_b = base pitch (distance along base circle between a point on a tooth and the corresponding point on the adjacent tooth), in.
ϕ = pressure angle

A simpler way to find the contact ratio is that defined by Eq. 7-12. A much simpler layout (Fig. 7-18) is needed to find the length of the path of action than that (Fig. 7-19) required to find the length of the arc of action.

The contact ratio is sometimes regarded roughly as "the average number of pairs of teeth in contact." A more precise interpretation is given in Table 7-3.

Designers usually prefer the minimum contact ratio to be 1.4 to insure smooth action between gears.

TABLE 7-3

Value of m_p	Interpretation
1	One pair of teeth in contact all the time.
$1 < m_p < 2$	One pair of teeth in contact $100(2/m_p - 1)\%$ of the time. Two pairs in contact the remaining time.
2	Two pairs in contact all the time.
$2 < m_p < 3$	Two pairs in contact $200(3/m_p - 1)\%$ of the time. Three pairs in contact the remaining time.
3	Three pairs in contact all the time.

Example 7-3. A pair of spur gears have teeth of 4 diametral pitch. The pressure angle is 20°, and the tooth addendum is to be $\frac{1}{4}$ in. The pitch diameters are 6 and 8 in. Determine the contact ratio.

Solution. The graphical portion of the solution is shown in Fig. 7-20.

1. The center distance is

$$C = \frac{D_p + D_g}{2}$$
$$= 7$$

The pitch radii of the pinion and gear are 3 and 4 in. respectively.

2. Lay out the line of centers, O_2O_3, and the pitch point, P, such that

$$O_2O_3 = C = 7 \text{ in.}$$
$$O_2P = R_p = 3 \text{ in.}$$

Point O_2 is the pinion's center of rotation.

3. Lay out the common tangent through P and perpendicular to the line of centers.

4. The problem does not state whether the pinion or the gear is the driver; nor does it specify the directions of rotation. However, the contact ratio does not depend on these considerations. It has been assumed, in this example, that the pinion is the driver and is turning clockwise.

5. Lay out the line of action through P. The angle between the line of action and the common tangent must be the pressure angle, ϕ. The proper position of the line of action is the position the common tangent would have if it were rotated about P through an angle ϕ in the direction of the driven gear's rotation. In this example, therefore, the position of the line of action is 20° ccw from the common tangent.

6. Lay out the addendum circle for each gear. The radius of the addendum circle is the sum of the pitch radius and the addendum.

7. Locate and label the points of initial and final contact, I and F.

8. Measure the length, Z, of the path of contact, IF. In the example, it measures 1.2 in.

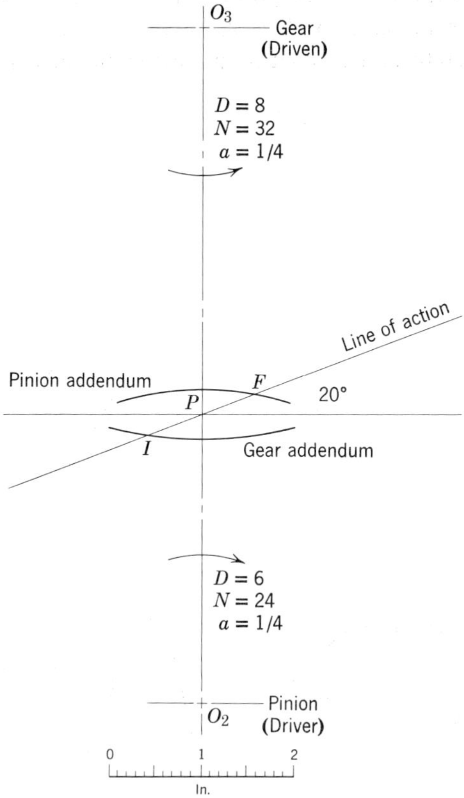

Fig. 7-20. Example problem 7-3.

9. Calculate the contact ratio, using Eqs. 7-13 and 7-12.

$$p_b = p \cos \phi$$
$$= (\pi/4) \cos 20°$$
$$= 0.738$$

$$m_p = \frac{Z}{p_b}$$
$$= \frac{1.2}{0.738}$$
$$= 1.63$$

7-9. Interference and Undercut

Interference is the mating of portions of tooth profiles which are not conjugate. In an involute tooth profile, the involute extends only to the base circle. Therefore, if contact occurs below the base line on a tooth's

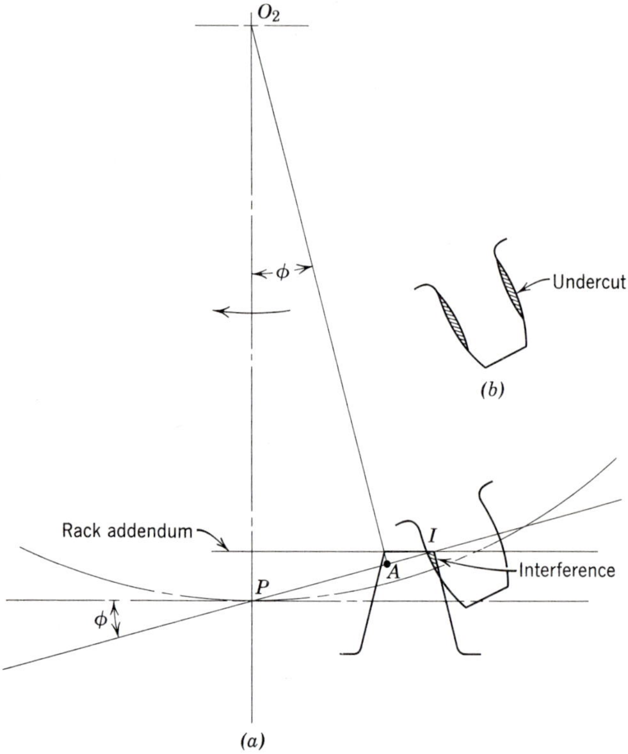

Fig. 7-21. Interference.

flank, rough action and binding will occur. The tip of the mating tooth will try to dig out the flank.

Interference is illustrated in Fig. 7-21*a*. The most serious cases of interference happen when a small pinion is meshing with a rack. In the illustration it has been assumed that the rack is the driven member, the pinion being the driver. The point of initial contact, *I*, is where the addendum line of the rack intersects the line of action. However, this point falls outside point *A*, where the line of action and base circle are tangent. This is evidence that contact between the teeth begins too soon, before the involute portion of the flank on the pinion tooth has come into position.

Contact takes place below the involute segment of the pinion tooth's profile. As the illustration shows, the teeth overlap, the tip of the rack tooth trying to dig out the flank of the pinion tooth.

To avoid interference, the initial point of contact should be at or inside A. Point A is called an *interference point*, for it is the limiting position of the initial point of contact. One way of avoiding interference would be to remove that portion of the rack tooth which extends above point A.

Now, if the rack happens to be a rack or hob cutter and the pinion is being cut, the cutter teeth will actually remove a portion of the flank of the pinion tooth, as shown in Fig. 7-21*b*. This is called *undercut*. This undercutting has the advantage of making interference impossible when the pinion is in use. But the weakening effect on the gear tooth produced by this undercutting is too great a price to pay.

Figure 7-22 illustrates the general graphical method for checking a gear

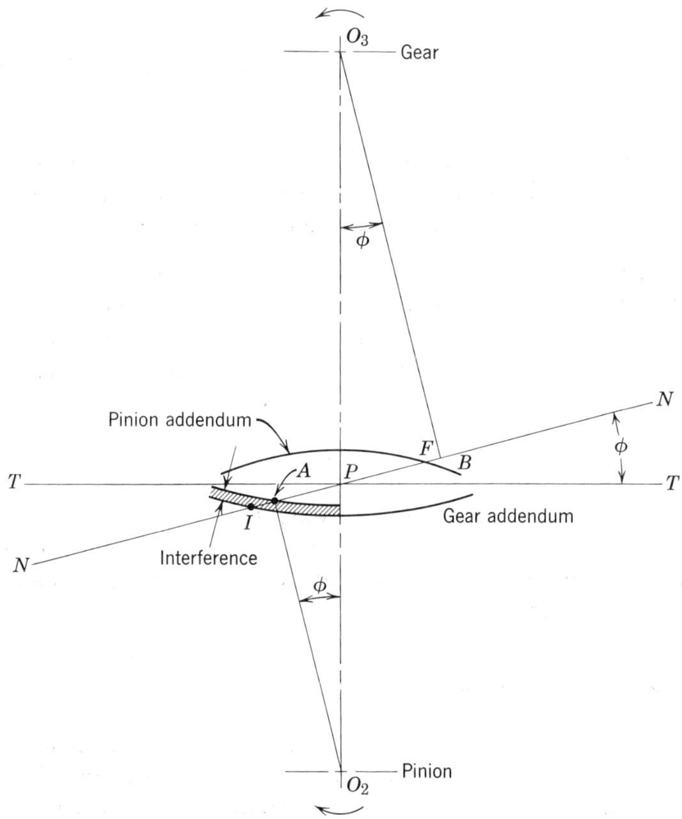

Fig. 7-22. Checking for interference.

pair for interference. It is assumed that the pressure angle, addendum, and pitch diameters are known. The directions of rotation need not be known; nor need we know which is the driver. These can be assumed if unknown. The following procedure can be used:

1. Lay out the line of centers, locating the centers of rotation and the pitch point.
2. Through the pitch point, P, draw the common tangent, T-T, perpendicular to the line of centers.
3. Through the pitch point draw the line of action, at angle ϕ with the common tangent, where ϕ is the pressure angle.
4. Locate the points A and B where the line of action is tangent to the base circles by drawing lines from the centers of rotation, each making angle ϕ with the line of centers. Where these lines (O_2A and O_3B) cross the line of action are the tangency points A and B (interference points).
5. Draw the addendum circles for the two gears and locate the points of initial and final contact, I and F.
6. If either I or F falls outside segment AB, there is interference. In Fig. 7-22 point I falls outside A. Hence there is interference.
7. To measure the amount of interference, draw an addendum circle for the gear through the interference point A. This addendum circle defines the maximum permissible length of the gear teeth. The radial dimension between the limiting and actual addendum circles is the interference.

Interference can be avoided by: (*a*) avoiding the use of small pinions, (*b*) undercutting the pinion flanks (weakens teeth), (*c*) relieving the gear teeth (reducing the addendum of the gear teeth by the amount of the interference or more), or (*d*) increasing the pressure angle (this may imply special cutting tools).

Example 7-4. A pair of spur gears have teeth with a pressure angle of 14° 30′ and an addendum of $\frac{1}{4}$ in. The pitch diameters are 5 and 10 in., and they have 20 and 40 teeth respectively. Check for interference.

Solution. The solution is shown in Fig. 7-23. The interference measures 0.06 in. Interference could be eliminated by reducing the addendum of the gear teeth by 0.06 in.

7-10. Standard Tooth Forms

A number of tooth forms have been standardized for interchangeable spur gears. To be interchangeable gears must have the same diametral pitch, the same pressure angle, same addendum, same dedendum, and same tooth thickness (half the circular pitch).

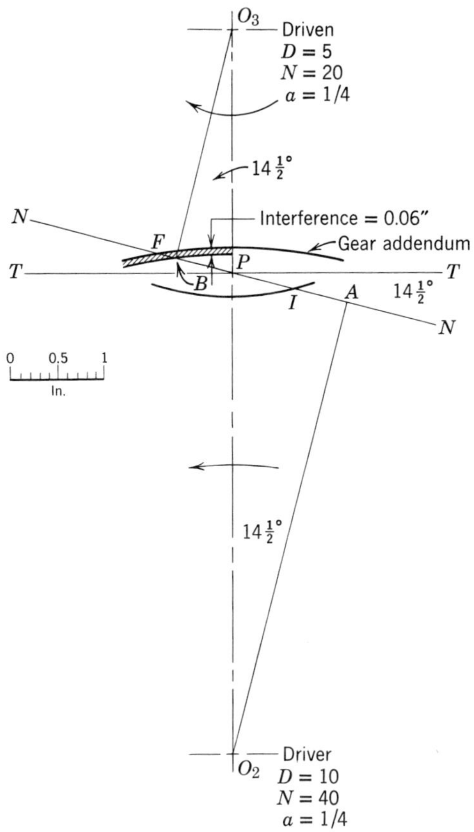

Fig. 7-23. Example problem 7-4.

For reasons of economy it is wise to use relatively few of the standard diametral pitches in order to keep to a minimum the stock of cutters required, unless there is good reason to do otherwise. The diametral pitches most often used are:

Standard Diametral Pitches Most Used

1, $1\frac{1}{2}$, 2, $2\frac{1}{2}$, 3, 4, 5, 6, 8,
10, 12, 16, 20, 24, 32, 40, 48, 64,
72, 96, 120, 160.

$14\frac{1}{2}°$ Composite system. This is a full-depth tooth cut by milling. A portion of the profile is involute, the rest being approximately cycloidal. This is the oldest tooth form and can be made in a machine shop possessing only a modest amount of equipment.

$14\frac{1}{2}°$ Full-depth involute system. Gears of this system are usually made by a generating process. Undercutting will occur when cutting a gear having fewer than 32 teeth, and it will become excessive for gears having under 22 teeth. For pinions with fewer than 20 teeth, the contact ratio will be under 1.4 no matter how large a gear is mating with it. However, if the pinion has at least 22 teeth, the contact ratio will be at least 1.4. The low pressure angle of this system results in lower bearing loads than for the 20° systems. This system is recommended where the pinion will have about 40 teeth or more.

20° Full-depth involute system. As with the previous system, gears of this type are made by generating. Although the larger pressure angle

TABLE 7-4

	System	
Dimension	$14\frac{1}{2}°$ composite $14\frac{1}{2}$ full-depth involute 20° full-depth involute	20° stub
Addendum	$1/P$	$0.8/P$
Dedendum (min.)	$1.157/P$	$1/P$
Full depth (min.)	$2.157/P$	$1.8/P$
Clearance (min.)	$0.157/P$	$0.2/P$

means heavier bearings, there are definite advantages. Undercutting is a problem only for gears having fewer than 18 teeth, becoming excessive for those with less than 14 teeth. If the pinion has 14 or more teeth, the contact ratio for a pair of these gears will be at least 1.4. This system is recommended for general use.

20° Stub involute system. Generally gears of this system are also made by a generating process. Here the problem of undercutting for most applications, is avoided. Undercut will occur only for pinions having fewer than 14 teeth. However, for meshing gears of equal size, the contact ratio is only 1.185 if each has 12 teeth, and 1.217 if each has 14 teeth. Thus the advantage gained from the viewpoint of undercutting is largely lost from the viewpoint of contact ratio. This system is recommended only for applications where an extremely small pinion must be used.

The standard tooth proportions for the above four types of tooth system are listed in Table 7-4.

20° Fellows stub tooth system. This system, devised by the Fellows Gear Shaper Company, is similar to the 20° Stub System except that the tooth proportions vary. Among the typical pitch designations in this system are: $\frac{3}{4}$, $\frac{4}{5}$, $\frac{5}{7}$, $\frac{6}{8}$, $\frac{7}{9}$, etc. The diametral pitch is the numerator of the fractional designation. The addendum is the reciprocal of the denominator, and the clearance is 0.25 divided by the denominator. For example, for the $\frac{5}{7}$ pitch tooth, the diametral pitch is 5, the addendum is $\frac{1}{7} = 0.1429$, and the clearance is $0.25/7 = 0.0357$.

20° Involute fine-pitch system. This is a tooth form for gears of 20 diametral pitch or finer. It is similar to the 20° full-depth involute system, but it provides greater clearance by a slight increase in the dedendum. The dedendum is made $1/P + 0.002$.

7-11. Helical Gears

Although in spur gears, the entire tooth width comes into contact with the mating tooth at once, helical gear teeth pick up the load gradually. Because of their smoother engagement and, further, because of their greater strength, helical gears can be used for higher speeds and heavier loads than spur gears of comparable size.

Terminology. In Fig. 7-24 a highly schematic drawing of a helical gear illustrates some of the basic terminology. The teeth here are shown straight and as lines. Their actual appearance can be seen in Figs. 7-2 and 7-5.

The *circular pitch in plane of rotation*, p, has the same significance as the circular pitch in spur gears. It is the distance between corresponding points on adjacent teeth measured on the pitch circle.

The *normal circular pitch*, p_n, is measured in a direction normal to the teeth, as shown in Fig. 7-24.

The *helix angle*, ψ, is the angle between an element on the helical tooth and the axis of rotation of the gear.

Basic relationships. Inspection of Fig. 7-24 justifies the following relationship:

$$p_n = p \cos \psi \tag{7-14}$$

The following equation defines *normal diametral pitch*:

$$P_n = \frac{\pi}{p_n} \tag{7-15}$$

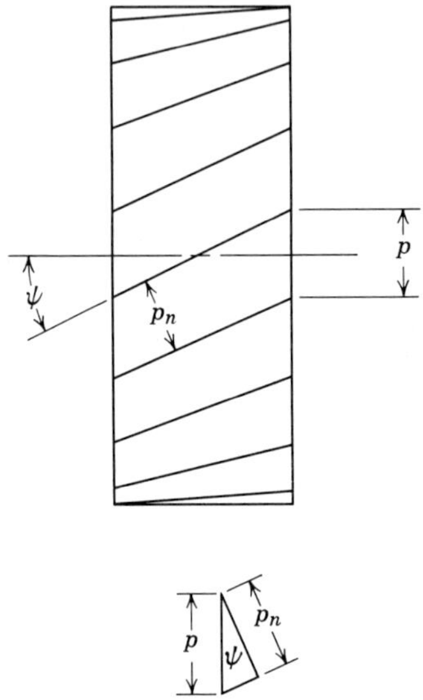

Fig. 7-24. Helical gear notation.

Substituting for p_n from Eq. 7-14 and in Eq. 7-15,

$$P_n = \frac{\pi}{p \cos \psi} \tag{7-15a}$$

As in spur gears,

$$pP = \pi \tag{7-3}$$

Substituting in Eq. 7-15*a* for *p* from Eq. 7-3,

$$P = P_n \cos \psi \tag{7-16}$$

where p = circular pitch in plane of rotation, in.
p_n = normal circular pitch, in.
P = diametral pitch, in.^{-1}
P_n = normal diametral pitch, in.^{-1}
ψ = helix angle

From Eq. 7-2,

$$N = PD$$

Substituting from Eq. 7-16 for *P*,

$$N = P_n D \cos \psi \tag{7-17}$$

From Eq. 7-5,

$$C = \frac{D_p}{2} + \frac{D_g}{2}$$

Substituting for D from Eq. 7-17,

$$C = \frac{N_p}{2P_n \cos \psi_p} + \frac{N_g}{2P_n \cos \psi_g} \tag{7-18}$$

As in spur gears, the speeds vary inversely as the numbers of teeth.

$$\frac{\omega_p}{\omega_g} = \frac{N_g}{N_p} \tag{7-8}$$

Substituting for N from Eq. 7-17,

$$\frac{\omega_p}{\omega_g} = \frac{D_g \cos \psi_g}{D_p \cos \psi_p} \tag{7-19}$$

Shaft and helix angles. When helical gears connect parallel shafts, they have the same helix angle, though they must be of opposite hand. If, for example, the pinion is right-handed, then the gear must be left-handed.

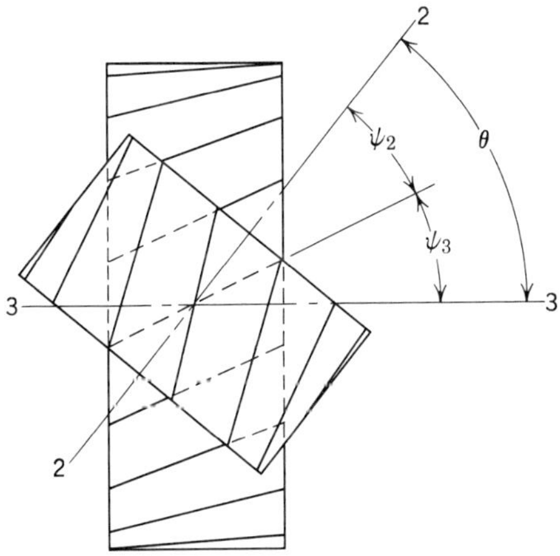

Fig. 7-25. Helical gears connecting skew shafts.

When helical gears connect skew axes, they may be of the same or of opposite hand. And their helix angles are, in general, not the same. The following two rules govern the relationship of shaft and helix angles.

1. If the gears are of the same hand, the shaft angle is the sum of the helix angles.
2. If the gears are of opposite hand, the shaft angle is the difference of the helix angles.

The first of these rules is illustrated in Fig. 7-25. Here the gears are of the same hand, both being left-handed. And the shaft angle, θ, is the sum of the helix angles, ψ_2, and ψ_3. Gears connecting parallel shafts are a special case of the second rule. They are of opposite hand; so the shaft angle must be the difference of the helix angles. But the shaft angle is 0°. Therefore the difference of the helix angles must be 0°, that is, the helix angles must be the same.

7-12. Worm Gears

Worm gear drives (Fig. 7-6) permit the transmission of power with a high speed ratio and use of a very compact unit. The double-throated design of Fig. 7-6*c* makes it possible to transmit heavy loads.

Worm gear drives ordinarily connect skew shafts which are at right angles, and usually the worm is the driver. The ratio of speeds is given by the equation

$$\frac{\omega_{\text{worm}}}{\omega_{\text{wheel}}} = \frac{N}{n} \tag{7-20}$$

where N = number of teeth on worm wheel

n = number of threads on worm: that is, if the worm is single-threaded, $n = 1$; if the worm is double threaded, $n = 2$; etc.

7-13. Bevel Gears

Bevel gear drives, Fig. 7-4, connect shafts that intersect. The drive can be either external or internal, depending on the shaft directions required.

Some of the terminology of bevel gears is shown in Fig. 7-26.

Kinematically bevel gears are equivalent to rolling cones. The pitch cones for an external drive are shown in Fig. 7-27*a* and for an internal drive in Fig. 7-27*b*. A relationship is needed between the shaft angle, pitch angles, and rotational speeds of the cones.

In Fig. 7-27*a*,

$$r_2 = OA \sin \alpha$$

and

$$r_3 = OA \sin \beta$$

Therefore,

$$\frac{r_2}{r_3} = \frac{\sin \alpha}{\sin \beta}$$

and

$$\sin \alpha = \frac{r_2}{r_3} \sin \beta$$

But, for the external drive,

$$\alpha + \beta = \theta \tag{7-21}$$

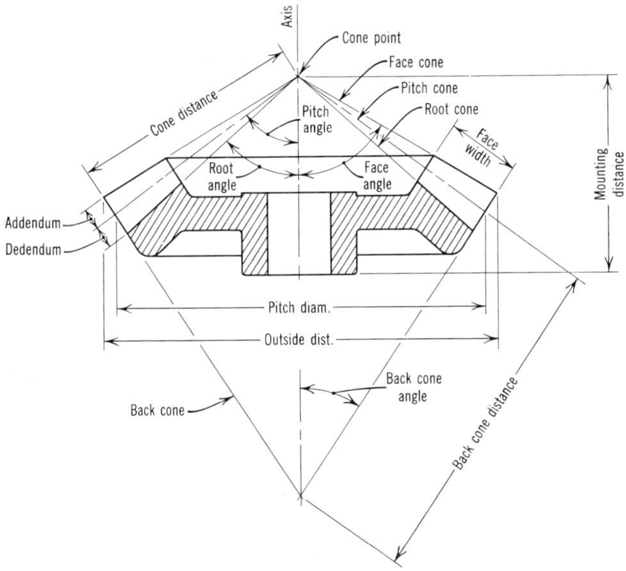

Fig. 7-26. Bevel gear terminology.

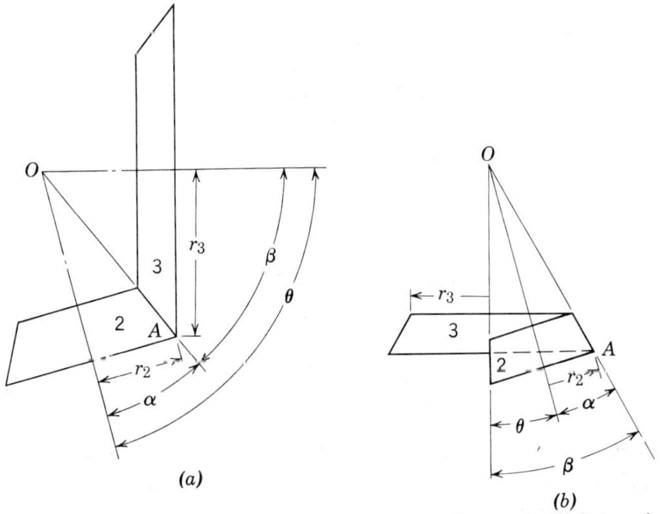

Fig. 7-27. Rolling pitch cones of bevel gears. (*a*) External contact. (*b*) Internal contact.

Therefore,
$$\sin \alpha = \frac{r_2}{r_3} \sin (\theta - \alpha)$$

or
$$\sin \alpha = \frac{r_2}{r_3} (\sin \theta \cos \alpha - \cos \theta \sin \alpha)$$

Dividing both sides by $\cos \alpha$,

$$\tan \alpha = \frac{r_2}{r_3} (\sin \theta - \cos \theta \tan \alpha)$$

Solving for $\tan \alpha$,

$$\tan \alpha = \frac{\sin \theta}{r_3/r_2 + \cos \theta}$$

The cones are in rolling contact. Therefore for point A, a point of contact between the cones,

$$V_A = r_2\omega_2$$

and
$$V_A = r_3\omega_3$$

Therefore,
$$\frac{r_3}{r_2} = \frac{\omega_2}{\omega_3}$$

Substituting in the equation for $\tan \alpha$,

$$\tan \alpha = \frac{\sin \theta}{\omega_2/\omega_3 + \cos \theta} \tag{7-22}$$

In a similar manner, the following equation can be derived:

$$\tan \beta = \frac{\sin \theta}{\omega_3/\omega_2 + \cos \theta} \tag{7-23}$$

For internal contact, Fig. 7-27*b*,

$$\beta - \alpha = \theta \tag{7-24}$$

$$\tan \alpha = \frac{-\sin \theta}{\omega_2/\omega_3 - \cos \theta} \tag{7-25}$$

$$\tan \beta = \frac{\sin \theta}{\omega_3/\omega_2 - \cos \theta} \tag{7-26}$$

Example 7-5. A pair of bevel gears having external contact are to connect shafts whose shaft angle is 80°. The speeds are 1800 rpm and 600 rpm. Find the pitch angles.

Solution. Let the pinion be gear 2 and the gear be gear 3. Then, using Eqs. 7-22 and 7-23,

$$\tan \alpha = \frac{\sin 80}{1800/600 + \cos 80}$$

$$\tan \alpha = 0.31031$$

$$\alpha = 17^\circ\ 14'\ 22''$$

$$\tan \beta = \frac{\sin 80}{600/1800 + \cos 80}$$

$$\tan \beta = 1.94250$$

$$\beta = 62^\circ\ 45'\ 38''$$

As a check, Eq. 7-21 can be used.

$$17^\circ\ 14'\ 22'' + 62^\circ\ 45'\ 38'' = 80^\circ \quad \textit{Checks.}$$

7-14. Gear Manufacture

Gears are fabricated by a large proportion of the techniques available for the production of metal and plastic parts. The methods used vary

TABLE 7-5

SOME GEAR MANUFACTURING METHODS

Method	Typical Applications
Punching	Toys and timing devices.
Die casting	For low cost gears where the gear is an integral part of another machine element such as a cam.
Cold drawing and extruding	For low cost pinions with smooth, work-hardened teeth. Used in devices where friction must be minimized.
Powder metallurgy	High volume, inexpensive gears for home appliances.
Plastic molding	For quiet operation at low speeds and light loads; used in film projectors and business machines.
Milling	This and other gear cutting processes are used where accuracy and power capacity are important. Milling is used for low production or for very large gears.
Generating	Usual method for fabricating metal gears.
Shaving	Finishing process for unhardened gears.
Grinding	Finishing process for hardened gears.

widely in cost, the type of material that can be formed, and the accuracy of the finished product. The gear designer has to weigh all these factors when deciding upon the method of manufacture. Table 7-5 summarizes some of the widely used gear manufacture methods.

In milling gears, a milling cutter whose shape is that of the tooth space is used. One space is cut and the gear indexed for the next cut. Milling

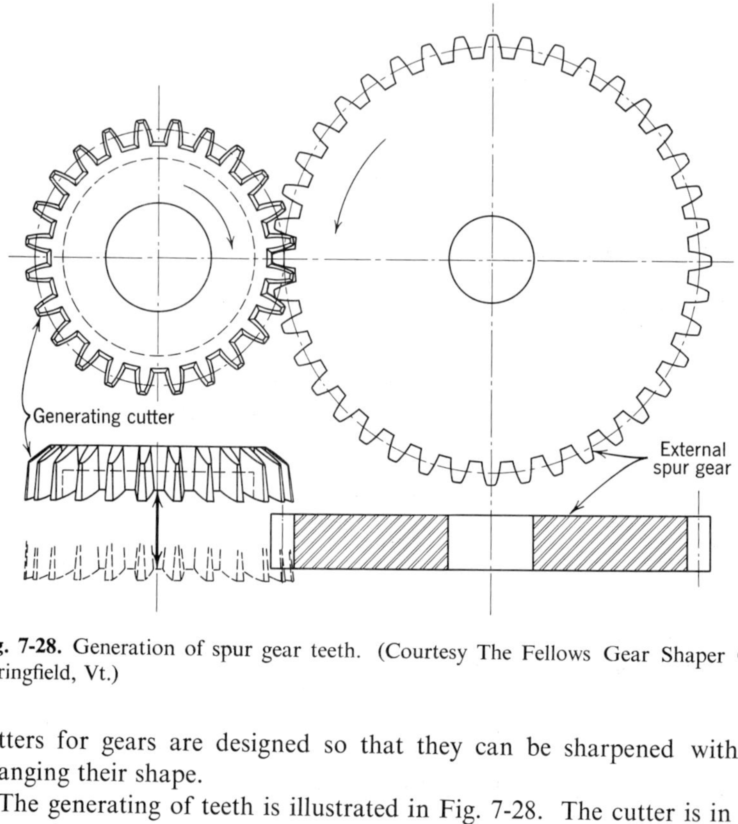

Fig. 7-28. Generation of spur gear teeth. (Courtesy The Fellows Gear Shaper Co., Springfield, Vt.)

cutters for gears are designed so that they can be sharpened without changing their shape.

The generating of teeth is illustrated in Fig. 7-28. The cutter is in the form of a spur gear except that its teeth are long enough to cut out the clearance spaces on the external spur gear being cut. As the cutter reciprocates, the cutter and gear roll together. For cutting helical gears, a helical cutter is used. This is illustrated in Fig. 7-29.

The fastest method of gear production is hobbing. Hobs can be used on standard milling machines or on special gear hobbing machines. The hob is shaped rather like a worm except that the threads are gashed to provide cutting edges. The hob is fed slowly across the face of the gear being cut while the hob and gear rotate at speeds inversely proportional to their

Fig. 7-29. Cutting helical gear teeth on a gear shaper. (Courtesy The Fellows Gear Shaper Company, Springfield, Vt.)

number of teeth. By the time the hob has completed its pass across the face of the gear, all the teeth have been cut.

REFERENCES

1. Fred H. Colvin and Frank A. Stanley, *Gear Cutting Practice*, McGraw-Hill Book Co., New York, 1950.
2. Darle W. Dudley, *Practical Gear Design*, McGraw-Hill Book Co., New York, 1954.
3. P. Houghton, *Gears: Spur, Helical, Bevel, and Worm*, The Technical Press, London, 1952.
4. Franklin D. Jones, *Gear Design Simplified*, 2nd ed., The Industrial Press, New York, 1940.

PROBLEMS

7-1. A spur gear has 28 teeth of 4 diametral pitch. Determine: (*a*) the pitch diameter, (*b*) the circular pitch.

7-2. A pair of spur gears connect shafts 7.5 in. apart and rotating at speeds of 1700 rpm and 430 rpm. Determine: (*a*) the pitch diameters, (*b*) the gear ratio.

7-3. Two meshing spur gears have teeth of 8 diametral pitch. The pinion has 22 teeth and the gear 47 teeth. Determine: (*a*) the pitch diameters, (*b*) the gear ratio, (*c*) the center distance, (*d*) the circular pitch.

7-4. A pinion rotating at 1200 rpm has 17 teeth. (*a*) How many teeth should there be on the mating gear if it is to rotate at approximately 750 rpm? (*b*) What will be the actual gear speed?

7-5. A pair of meshing spur gears have 16 and 74 teeth of 3 diametral pitch. What is the center distance?

7-6. A pair of spur gears are to connect shafts exactly 11 in. apart. The pinion speed is 1755 rpm, and the gear speed is to be as close to 550 rpm as possible. The pinion shall have no fewer than 18 teeth, and the diametral pitch shall be 2, $2\frac{1}{2}$, 3, 4, 5, or 6. Determine the pitch diameters, diametral pitch, and numbers of teeth for the gear pair that best meets the above specifications.

7-7. Derive an equation similar to Eq. 7-6, but for internal contact.

7-8. Determine the base circle diameter for a 17 in. pitch diameter spur gear having 20° stub-involute teeth.

7-9. Determine the contact ratio for a pair of spur gears having 20° full-depth teeth of 2 diametral pitch. The pinion has 24 teeth and the gear 41 teeth.

7-10. Determine the contact ratio for a pair of spur gears having $14\frac{1}{2}°$ full-depth involute teeth of 3 diametral pitch. The pinion and gear each have 40 teeth.

7-11. A spur gear has 28 teeth of 6 diametral pitch. They are $14\frac{1}{2}°$ full-depth involute. Determine: (*a*) pitch diameter, (*b*) circular pitch, (*c*) addendum, (*d*) minimum dedendum, (*e*) minimum clearance, (*f*) minimum whole depth, (*g*) base circle radius.

7-12. A spur gear has 16 teeth of the 20° stub-involute system. The circular pitch is 0.7854 in. Determine: (*a*) the diametral pitch, (*b*) the pitch diameter, (*c*) the addendum, (*d*) the minimum dedendum, (*e*) the base circle diameter.

7-13. A spur pinion having 14 teeth of 20° full-depth involute form and $2\frac{1}{2}$ diametral pitch is meshing with a 34 tooth gear. Determine the amount, if any, of interference.

7-14. A spur pinion having 26 teeth of $14\frac{1}{2}°$ full-depth form meshes with a rack. The diametral pitch is 3. Determine the amount of interference, if any.

7-15. A helical pinion has 14 teeth of 6 diametral pitch. The helix angle is 22°. Determine: (*a*) the circular pitch, (*b*) the normal circular pitch, (*c*) the normal diametral pitch, (*d*) the pitch diameter.

7-16. A pair of helical gears connect skew shafts. The normal diametral pitch is 4. The pinion has 22 teeth and the gear 38 teeth. The helix angle of the pinion is 20° and that of the gear 23°. Both gears are right-handed. Determine: (*a*) the center distance, (*b*) the shaft angle.

7-17. A pair of helical gears are to connect parallel shafts 5.1 in. apart. The pinion has 16 teeth and the gear 32 teeth of 5 normal diametral pitch. Determine the helix angle.

7-18. A double-threaded worm running at 1800 rpm drives a worm wheel which is to run at about 97 rpm. Determine: (*a*) the number of teeth on the wheel, (*b*) the actual output speed.

7-19. A single threaded worm drives a worm wheel having 73 teeth. If the output speed is to be 14 rpm, what should be the input speed?

7-20. Bevel gears having external contact connect shafts at 73°. The pinion speed is 700 rpm and the gear speed is 300 rpm. Determine the pitch angles.

7-21. Bevel gears having internal contact connect shafts at 19° and run at speeds of 1200 rpm and 750 rpm. Find the pitch angles.

8 Mechanism Trains

8-1. Introduction

When applying mechanisms to machines or instruments, diverse goals can be realized simultaneously by using differing mechanisms together in a *mechanism train.* Simpler units such as gear trains, belt drives, cams, linkages, and hydraulic systems may be incorporated in a single machine. A type of output motion difficult to achieve with any single mechanism often can be achieved easily by combining mechanisms. Similarly, a required speed ratio or a mechanical advantage can often be obtained more

Fig. 8-1. Shaft mounted speed reducer. (Courtesy The American Pulley Co., Philadelphia, Pa.)

compactly by a train incorporating a number of simpler mechanisms than by a single mechanism alone.

Figure 8-1 illustrates a belt-driven, shaft-mounted, speed reduction unit. The belt drive and the gear reduction unit are its two simpler components. The gear train provides a large speed reduction in a small space. The belt drive contributes to the speed ratio, makes possible the economical connection of shafts relatively far apart, and protects the motor in case the load stalls. In the event of stalling, the belts will simply slip.

8-2. Gear Trains

When a series of meshing gears transmits motion from one shaft to another, or between a shaft and a slide, the series is called a *gear train*. Even a single pair of gears is, strictly speaking, a gear train, though the term usually suggests that there are three or more moving members.

Classification. The system of classification below groups the various types of gear trains into convenient categories.

I. Ordinary trains
  a. Simple
  b. Compound
    (1) Nonreverted
    (2) Reverted

II. Epicyclic trains
  a. One fixed gear
  b. No fixed gear

In an *ordinary gear train*, all gear axes remain fixed relative to the frame, as in Fig. 8-2. If, in addition, each gear is on a separate shaft, the train is said to be *simple* as well. The ordinary gear train in Fig. 8-2 is also simple, each of its gears being on a separate shaft.

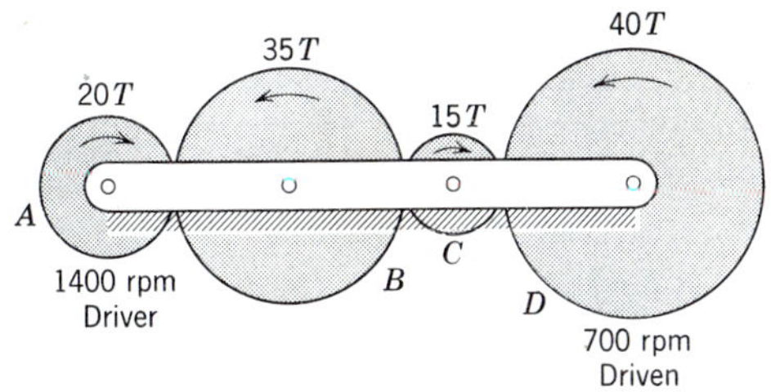

Fig. 8-2. Ordinary gear train.

If, on the other hand, at least one shaft has two or more gears keyed to it, the train is *compound.* In the four-gear compound gear train of Fig. 8-3, gear A is the driver, driving B. Gear C, rotating at the same speed as B, drives the last driven gear D.

Compound gear trains are in turn subdivided into two groups, nonreverted and reverted. A *reverted* train is one so designed that the input and output shafts line up. In Fig. 8-4 the shafts carrying gears A and D, the driver and last driven gear, line up.

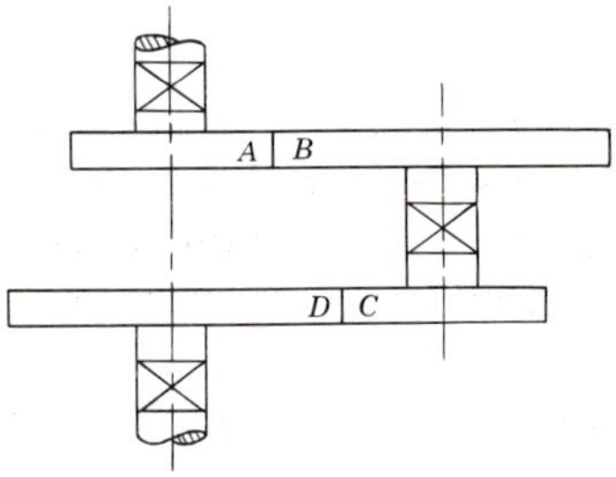

Fig. 8-3. Compound gear train.

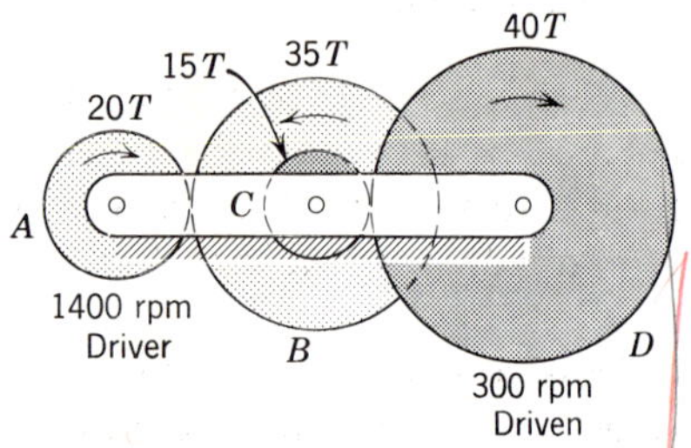

Fig. 8-4. Reverted gear train.

If at least one of the gear axes is in motion relative to the frame, the train is *epicyclic.* The distinction between ordinary and epicyclic trains is shown in Fig. 8-5. Here a gear assembly is used first as an ordinary gear train by fixing the arm, and then as an epicyclic train by releasing the arm and fixing gear B. In the epicyclic train of Fig. 8-5b the arm, or *planet carrier*, carries the *planet* gear C around the *sun* gear B. The words "sun" and "planet" in this context are used because of the analogy between the motion of the planet gear around the sun gear and the motion of a planet around the sun.

An epicyclic train having no fixed gears is shown in Fig. 8-6. The epicyclic train itself consists of gears D, E, F, G, and H, and the arm. Gears A and B drive the epicyclic train. Epicyclic gear trains in general have two degrees of freedom; this is, it is necessary to have two input motions to the train in order to get a controlled output. In this instance gear A drives gear C, thus controlling the arm motion, and B drives D. Two definite input motions having been provided, the motion of the output shaft is controlled. Fixing a gear of an epicyclic gear train is equivalent to providing one of the two necessary input motions.

The familiar automobile differential in Fig. 8-7 is another illustration of an epicyclic unit with no fixed gears. It functions as an ordinary gear train when the car is traveling on a straight path, but when the car is rounding a curve or making a turn, it functions as an epicyclic train. In

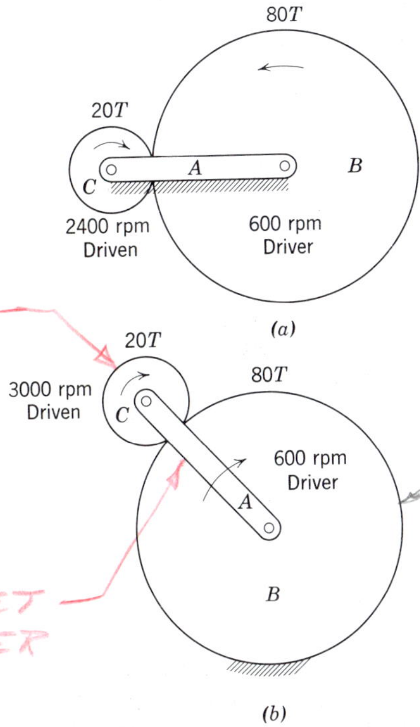

Fig. 8-5. Ordinary and epicyclic trains.

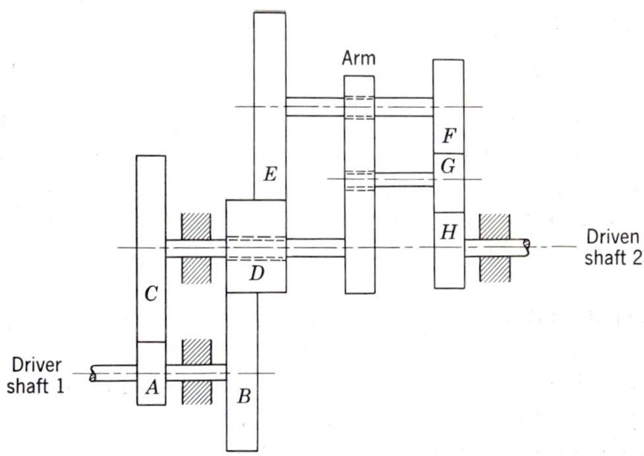

Fig. 8-6. Epicyclic gear train with no fixed gear.

the latter case, it is necessary that the axles travel at different speeds. The differential controls the relative speeds of the two axles: the sum of the axle speeds is always twice the ring gear speed.

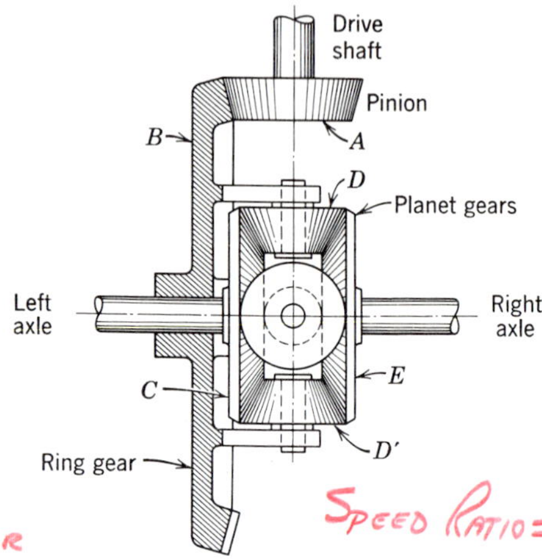

Fig. 8-7. Automobile differential.

Ratios. The ratio of the angular velocity of the last driven gear to that of the driving gear is called the *train value*. The ratio of the driver's velocity to that of the last driven train member is the *speed ratio*. Train value and speed ratio are reciprocal quantities. For the ordinary gear train of Fig. 8-2, the train value is 0.5, and the speed ratio is 2.0. The train value of the epicyclic unit in Fig. 8-5*b* is 5.

Train values and speed ratios can be specified as either plus or minus. A positive value indicates that the driving and driven gears are turning in the same direction, a negative value that their directions are opposite. The sign of a train value, or of a speed ratio, by itself does not indicate the direction of either the driving or the driven gear. It merely indicates whether their directions are the same or opposite. Often in gear train calculations, the sign of the ratio is ignored during the calculations. It can be specified later by inspecting the train.

Direction of rotation. The direction of a gear's rotation is conventionally specified as clockwise or counterclockwise. However, these terms are ambiguous unless the line of sight is specified as well. In Fig. 8-8 the gear is rotating in a counterclockwise direction for an observer looking from *A*, but clockwise for an observer looking from *B*.

Since, in trigonometry, positive angles are those measured in the counterclockwise direction, counterclockwise rotation will be regarded as positive and clockwise as negative. This convention will be adhered to in this chapter. However, there is seldom any harm in using the opposite convention so long as the convention adopted is explicitly stated in any written work and adhered to consistently.

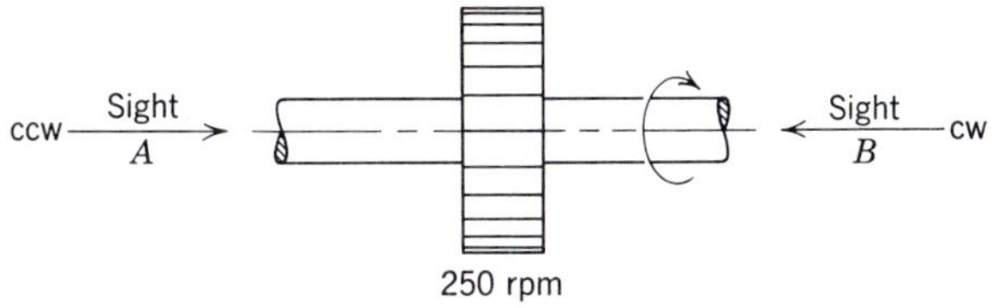

Fig. 8-8. Direction of rotation.

When the axes of a gear train are not parallel, there is no single line of sight that applies to the whole train. In this case it is better to dispense with the words "counterclockwise" and "clockwise" and to indicate the directions of rotation with arrows on a sketch of the train. This has been done in Fig. 8-9.

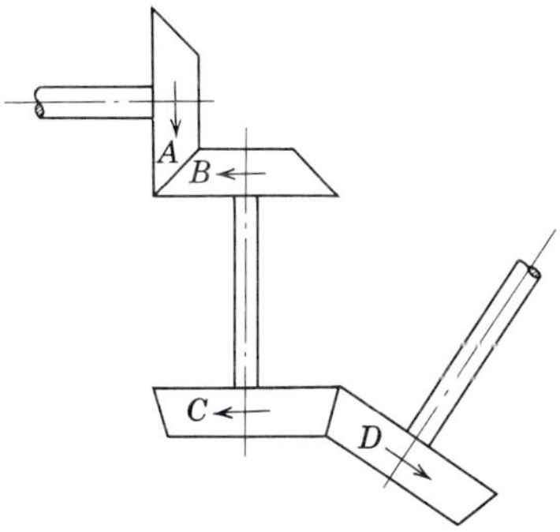

Fig. 8-9. Direction of rotation.

8-3. Ordinary Gear Trains

Simple. For the simple gear train in Fig. 8-10 the train value is, by definition, ω_{E}/ω_{A}. Because the speeds vary inversely as to the numbers of

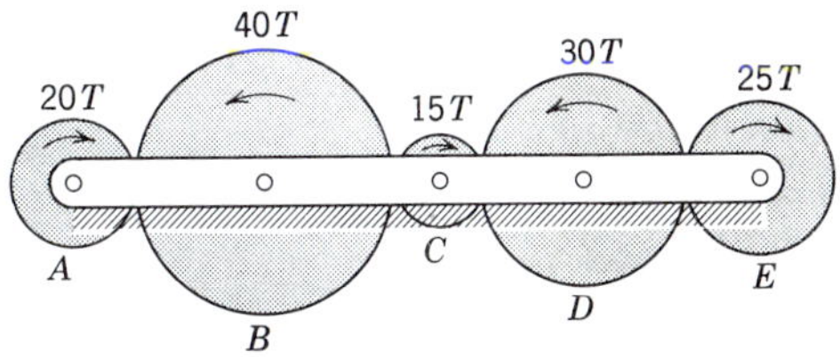

Fig. 8-10. Ordinary gear train.

teeth for any pair of meshing gears (ignoring signs):

$$\frac{\omega_B}{\omega_A} = \frac{N_A}{N_B}$$

$$\frac{\omega_C}{\omega_B} = \frac{N_B}{N_C}$$

$$\frac{\omega_D}{\omega_C} = \frac{N_C}{N_D}$$

$$\frac{\omega_E}{\omega_D} = \frac{N_D}{N_E}$$

Making the product of the left members of these equations equal to the product of the right members, there results

$$\frac{\omega_B}{\omega_A} \cdot \frac{\omega_C}{\omega_B} \cdot \frac{\omega_D}{\omega_C} \cdot \frac{\omega_E}{\omega_D} = \frac{N_A}{N_B} \cdot \frac{N_B}{N_C} \cdot \frac{N_C}{N_D} \cdot \frac{N_D}{N_E}$$

After cancellations, this reduces to

$$\frac{\omega_E}{\omega_A} = \frac{N_A}{N_E}$$

This same method of analysis can be extended to any simple gear train, and the result will always be the same:

$$\text{Train value} = \pm \frac{\text{Number of teeth on driver}}{\text{Number of teeth on last driven}} \tag{8-1}$$

Whether the train value will be positive or negative depends on the relative directions of the driving and last driven gears. The sign of the train value can be determined by inspection of a sketch of the train.

The *idlers* between the driving and last driven gears have no effect on the numerical value of the train value, but they serve other purposes. To connect shafts some distance apart without using excessively large gears, idlers are employed. Inserting an idler between gears in a simple gear train will reverse the direction of the last driven gear. And, where a single power source is to drive a series of shafts which are to turn in precise synchronization, a simple gear train can be used to advantage.

Example 8-1. If the angular velocity of the gear A in Fig. 8-10 is 1600 rpm clockwise, what will be the angular velocity of gear E?

Solution. From an inspection of the figure it is evident that gears A and E will turn in the same direction. Applying Eq. 8-1, but ignoring signs,

$$\frac{\omega_E}{\omega_A} = \frac{N_A}{N_E}$$

$$\frac{\omega_E}{1600} = \frac{20}{25}$$

$$\omega_E = 1280 \text{ rpm}$$

Since gears E and A turn in the same direction, the angular velocity of E is 1280 rpm clockwise.

Compound trains. For the compound train of Fig. 8-11, the train value is, by definition, ω_F/ω_A. For each pair of meshing gears, the speeds vary

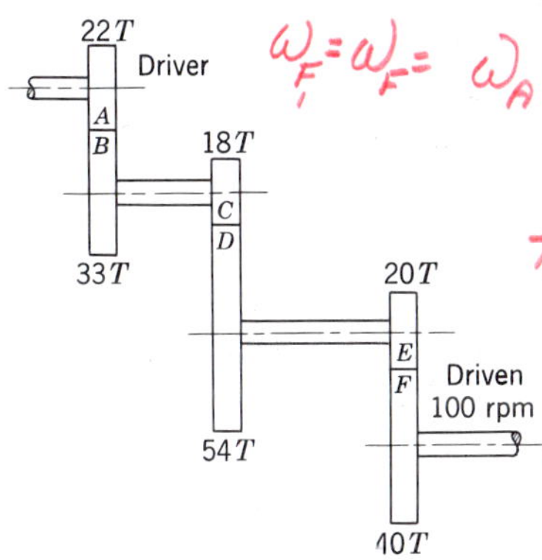

Fig. 8-11. Compound gear train.

inversely as to the numbers of teeth. Therefore,

$$\frac{\omega_B}{\omega_A} = \frac{N_A}{N_B}$$

$$\frac{\omega_D}{\omega_C} = \frac{N_C}{N_D}$$

$$\frac{\omega_F}{\omega_E} = \frac{N_E}{N_F}$$

Making the product of the left terms equal to the product of the right terms, there results

$$\frac{\omega_B}{\omega_A} \cdot \frac{\omega_D}{\omega_C} \cdot \frac{\omega_F}{\omega_E} = \frac{N_A}{N_B} \cdot \frac{N_C}{N_D} \cdot \frac{N_E}{N_F}$$

But the speeds of B and C are equal since they are keyed to the same shaft. Similarly, the speeds of D and E are equal. Therefore, after cancellations, the above equation reduces to

$$\frac{\omega_F}{\omega_A} = \frac{N_A \cdot N_C \cdot N_E}{N_B \cdot N_D \cdot N_F}$$

Inspection of this equation reveals the fact that the numerator of the right-hand side is the product of the numbers of teeth on the drivers, while the denominator is the product of the numbers of teeth on the drivens. This same result will occur in analyzing any ordinary, compound gear train. Thus, for a compound train,

$$\text{Train value} = \pm \frac{\text{Product of numbers of teeth on drivers}}{\text{Product of numbers of teeth on drivens}} \qquad (8\text{-}2)$$

Equation 8-2 applies, actually, to all ordinary gear trains, for it may be applied to simple gear trains as well as to compound ones. When applied to a simple gear train, each idler will appear both in the denominator as a driven gear and in the numerator as a driver. The numbers of teeth on the idlers will, therefore, cancel out.

Example 8-2. Determine the train value for the compound gear train of Fig. 8-11. What must be the angular velocity of gear A if the angular velocity of gear F is to be 100 rpm clockwise?

Solution. Applying Eq. 8-2, but ignoring signs,

$$\frac{\omega_F}{\omega_A} = \frac{22 \cdot 18 \cdot 20}{33 \cdot 54 \cdot 40}$$

$$\frac{\omega_F}{\omega_A} = \frac{1}{9}$$

Since an inspection of the train reveals that gears A and F must turn in opposite directions, the train value is $-\frac{1}{9}$. For gear F to turn at 100 rpm clockwise, then, it is necessary that gear A turn at 900 rpm counterclockwise.

For the reverted train of Fig. 8-12, it is necessary that the center distance for each pair of meshing gears be the same. For spur gears, Eq. 7-6 can be applied to each pair of meshing gears.

$$2PC = N_A + N_B$$

$$2P'C = N_C + N_D$$

where P = diametral pitch for gears A and B
P' = diametral pitch for gears C and D

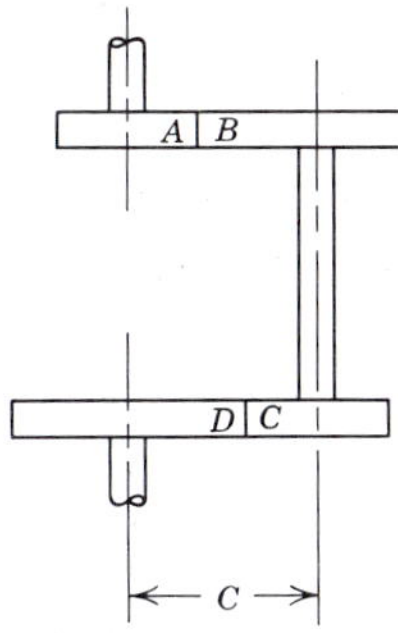

Fig. 8-12. Reverted gear train.

Therefore,

$$\frac{N_A + N_B}{N_C + N_D} = \frac{P}{P'} \tag{8-3}$$

For helical gears, Eq. 7-18 can be applied to each meshing pair.

$$2P_n C \cos \psi = N_A + N_B$$

$$2P'_n C \cos \psi' = N_C + N_D$$

where the primed diametral pitch and helix angle apply to gears C and D. Combining these two equations, there results

$$\frac{N_A + N_B}{N_C + N_D} = \frac{P_n \cos \psi}{P'_n \cos \psi'} \tag{8-4}$$

A comparison of Eqs. 8-3 and 8-4 reveals that in designing reverted gear trains, helical gears provide considerably more flexibility in design than spur gears. The helix angles can be varied at will, within practical limits, by the designer in order to find a combination of gears that can be used in the reverted train and that will satisfy the other requirements as well.

8-4. Transmissions

When a machine will be operated at any one of several output speeds, a multiple-speed gearbox, or transmission, may be incorporated as a component part. Machine tools and motor vehicles are familiar instances of the need for transmissions. The speed of the output shaft of the transmission can be varied by sliding gears in and out of contact or by connecting gears in continual mesh to shafts by means of clutches.

It is common practice in transmission design to arrange the magnitudes of the output speeds in a prescribed pattern, such as a geometric progression; that is, such that the ratio between successive speeds is constant.

Example 8-3. The lowest in a series of six speeds is to be 100 rpm. Each higher speed is to be 20% greater than the previous one. What are the speeds of the series?

Solution. Since each speed is to be 20% greater than the previous one, each speed is 1.20 times the previous one. The solution is shown in Table 8-1.

TABLE 8-1

Speed	Magnitude
1	100
2	$1.2 \times 100 = 120$
3	$1.2 \times 120 = 144$
4	$1.2 \times 144 = 173$
5	$1.2 \times 173 = 207$
6	$1.2 \times 207 = 249$

The process illustrated in the above example can be generalized as follows:

Let ω = the lowest speed in the series
k = multiplication factor (in the example it was 1.20).
ω_n = the nth speed in the series

Then

$$\omega_1 = \omega$$

$$\omega_2 = k(\omega) = k\omega$$

$$\omega_3 = k(k\omega) = k^2\omega$$

$$\omega_4 = k(k^2\omega) = k^3\omega$$

or, in general,

$$\omega_n = k^{n-1}\omega \tag{8-5}$$

Example 8-4. It is required to determine a series of eight speeds ranging from 100 rpm to 1000 rpm such that they form a geometric progression. Specify each speed to three significant figures.

Solution

$$\omega = 100$$

$$\omega_n = 1000$$

$$n = 8$$

Solving Eq. 8-5 for k,

$$k = \sqrt[n-1]{\frac{\omega_n}{\omega}}$$

$$= \sqrt[7]{10} = 1.3895$$

$$\omega_1 = \omega \quad = 100$$

$$\omega_2 = k\omega \;\;= (1.3895)(100) = 139$$
$$\omega_3 = k^2\omega = (1.3895)^2(100) = 193$$
$$\omega_4 = k^3\omega = (1.3895)^3(100) = 268$$
$$\omega_5 = k^4\omega = (1.3895)^4(100) = 373$$
$$\omega_6 = k^5\omega = (1.3895)^5(100) = 518$$
$$\omega_7 = k^6\omega = (1.3895)^6(100) = 720$$
$$\omega_8 = k^7\omega = \omega_n = 1000$$

The transmission designer's task—to provide a pattern series of output speeds—is complicated by the limitations of space and the demands of economy. The desired speed series must be achieved with a minimum number of gears and shafts. His freedom is further limited by the necessity of selecting tooth sizes capable of carrying the loads. Inevitably, then, a certain amount of trial and error is involved, but much needless labor can be saved by applying a systematic approach to the problem, such as those outlined in articles by Hall (Ref. 6) and by Dodge (Ref. 4). Two examples of successful transmission design will now be given as illustrations.

A compact, nine-speed transmission utilizing ten gears and three shafts (discussed by Hall in Ref. 6) is shown in Fig. 8-13. Gears *A*, *B*, *C* slide together as a unit on shaft 1, making possible three different pairs

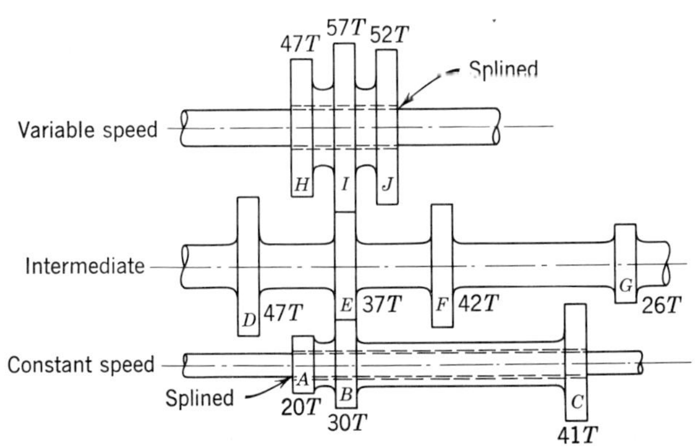

Fig. 8-13. Nine-speed transmission.

connecting shafts 1 and 2: *A* and *D*, or *B* and *E*, or *C* and *G*. Gears *H*, *I*, *J* slide as a unit on shaft 3, permitting shafts 2 and 3 to be connected by *D* and *H*, or *E* and *I*, or *F* and *J*. Consequently, nine possible combinations are available. Listed in Table 8-2 are the nine possible train values for this transmission.

TABLE 8-2

Speed No.	Gears in Mesh: Shafts 1 & 2	Gears in Mesh: Shafts 2 & 3	Train Value ω_3/ω_1
1	*A* & *D*	*E* & *I*	$\frac{20 \times 37}{47 \times 57} = 0.276$
2	*A* & *D*	*F* & *J*	$\frac{20 \times 42}{47 \times 52} = 0.344$
3	*A* & *D*	*D* & *H*	$\frac{20}{47} = 0.426$
4	*B* & *E*	*E* & *I*	$\frac{30}{57} = 0.526$
5	*B* & *E*	*F* & *J*	$\frac{30 \times 42}{37 \times 52} = 0.655$
6	*B* & *E*	*D* & *H*	$\frac{30 \times 47}{37 \times 47} = 0.811$
7	*C* & *G*	*E* & *I*	$\frac{41 \times 37}{26 \times 57} = 1.024$
8	*C* & *G*	*F* & *J*	$\frac{41 \times 42}{26 \times 52} = 1.274$
9	*C* & *G*	*D* & *H*	$\frac{41 \times 47}{26 \times 47} = 1.577$

A transmission of the type used in manually shifted motor vehicles is shown in Fig. 8-14. The following design features should be noted:

(*a*) Gear *A* is connected to the main clutch and operates at motor speed. It can turn freely on propeller shaft *P*.

(*b*) Gear *B* is always in mesh with *A*; and, consequently, countershaft *R* is always rotating as long as *A* is rotating.

(*c*) Gear *D* is continually in mesh with gear *C* and can turn freely on shaft *P*.

(*d*) Clutch *J* is keyed to shaft *P* but can slide axially. When moved to the left, it engages a clutch face on gear *A*, locking *A* and shaft *P* together. When moved to the right, it engages a clutch face on gear *D*, locking *D* and shaft *P* together.

(*e*) Gear *F* is keyed to shaft *P* but can slide axially along it. It can be meshed either with gear *E* on the countershaft or with idler gear *H*.

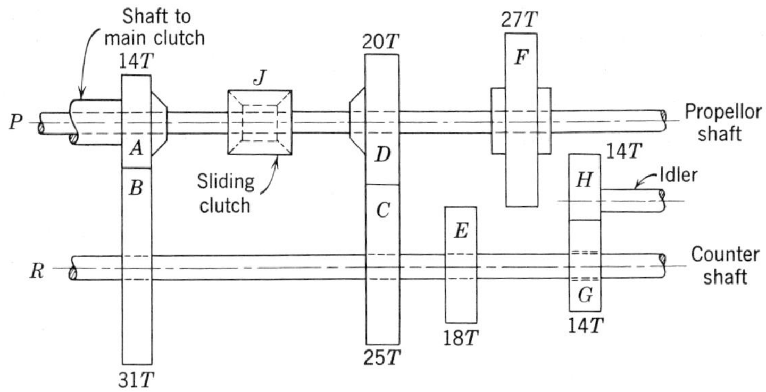

Fig. 8-14. Three-speed automotive transmission.

This transmission is operated by a single shift lever which has five positions: neutral, low, second, high, and reverse. The operation of the transmission is as follows:

(*a*) *Neutral*—The transmission is shown in neutral in the drawing. Clutch J and gear F are disengaged. Shaft P is unconnected to the rest of the transmission.

(*b*) *Low speed*—As the shift lever is moved into low position, it causes gear F to move into engagement with gear E. The power path is then $A—B—E—F$. The train value is

$$\frac{\omega_P}{\omega_A} = \frac{A \times E}{B \times F} = \frac{14 \times 18}{31 \times 27} = \frac{1}{3.32}$$

(*c*) *Second speed*—Gear F is moved out of engagement as the shift lever moves into neutral. As it moves to second, clutch J is moved to the right and into engagement with the clutch face on gear D. The power path is $A—B—C—D$. The train value is

$$\frac{\omega_P}{\omega_A} = \frac{A \times C}{B \times D} = \frac{14 \times 25}{31 \times 20} = \frac{1}{1.77}$$

(*d*) *High speed*—Clutch J is disengaged as the shift lever moves into neutral. And as the lever moves into high, the clutch continues to move to the left and into engagement with the clutch face on gear A. Gear A and shaft P are thus rotating together and at engine speed. The train value is 1.

(*e*) *Reverse*—Gear F is moved to the right and into engagement with idler gear H. The idler is necessary to reverse the direction of shaft P. The power path is $A—B—G—H—F$. The train value is

$$\frac{\omega_P}{\omega_A} = -\frac{A \times G}{B \times F} = -\frac{14 \times 14}{31 \times 27} = -\frac{1}{4.27}$$

A ten-speed transmission is shown in Fig. 8-15. Designed primarily for heavy duty, over-the-highway equipment, it can be shifted by a single lever. A selector handle on the shift lever permits using a basic five-speed pattern twice, the first time with an auxiliary section in low speed, and the second time with the auxiliary section in high. In addition, there are two reverse speeds.

Fig. 8-15. Ten-speed, heavy-duty transmission. (Courtesy Fuller Manufacturing Co., Kalamazoo, Mich.)

8-5. Epicyclic Trains

An epicyclic train has been defined as one in which there is relative motion of the gear axes. The importance of this type of train is steadily increasing. For a given number of gears, an epicyclic arrangement can provide a higher speed ratio than is obtainable from the same gears arranged in an ordinary train. Or, for a given ratio, an epicyclic unit can be considerably more compact, particularly when annular gears are employed. Epicyclic trains are finding important applications in transmissions, computing devices, and in many other ways.

Although epicyclic trains seem more difficult to analyze than ordinary trains because of the more complex motions, comparatively simple methods of analysis are available which do not require the ability to visualize accurately the epicyclic motions. The capacity to visualize their operation will develop as experience is gained in their analysis.

A simple epicyclic gear train is shown in Fig. 8-16. The sun gear B is stationary while the planet P, half its size, rolls around the sun, carried by the arm or planet carrier. The planet is shown in the drawing at four successive positions, and four points have been marked on its pitch circumference to make following its motion easier.

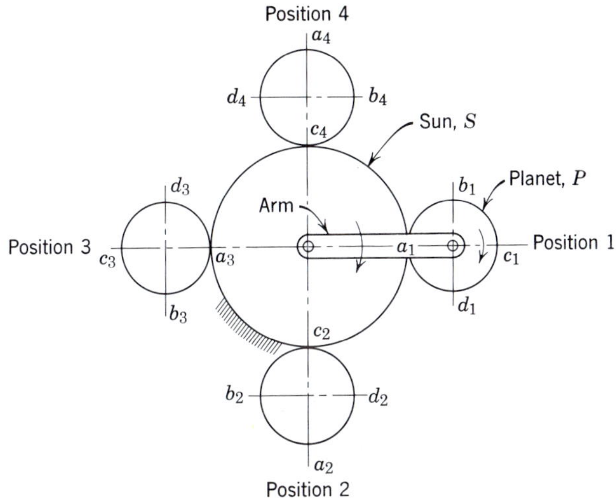

Fig. 8-16. Epicyclic train.

When the arm moves from position 1 to 2, the planet moves around the sun gear. And, since the planet is half the size of the sun, half the circumference of the planet corresponds to quarter the circumference of the sun. Thus point c rolls into contact with the sun gear at position 2. Relative to an observer on the arm, the planet has made half a revolution clockwise; but relative to the frame, the planet has made three fourths of a revolution clockwise. The same behavior can be observed between any two successive positions of the arm. As the arm makes a full turn clockwise, the planet moves two revolutions clockwise relative to the arm but three revolutions relative to the frame. The reader should examine Fig. 8-16 and convince himself that this is the case.

In general, for bodies in plane motion only,

$$\omega_3 = \omega_2 + \omega_{3/2} \tag{8-6}$$

where ω_3 = angular velocity of link 3 relative to frame
ω_2 = angular velocity of link 2 relative to frame
$\omega_{3/2}$ = angular velocity of link 3 relative to link 2

Applying this equation to the planetary gear train of Fig. 8-16, and taking the angular velocity of the arm to be +1 rpm,

$$\begin{aligned}\omega_P &= \omega_A + \omega_{P/A} \\ &= (1) + (2) \\ &= 3 \text{ rpm}\end{aligned}$$

It is convenient in epicyclic gear train analysis to write the relative angular velocity equation for each gear in terms of the planet carrier. Equation 8-6 can then be written as

$$\omega_G = \omega_A + \omega_{G/A} \tag{8-7}$$

where the subscript G indicates any gear under discussion and A indicates the arm or planet carrier.

8-6. Analysis of Epicyclic Gear Trains

Trains with one fixed gear. For this type of planetary train, the following three-step procedure is convenient.

1. Lock the train and rotate it +1 revolutions. Then each member of the train rotates +1 revolution.
2. Lock the arm, release the fixed gear, and turn this gear −1 revolutions. Record the number of revolutions made by each of the other gears. In this step the train has been made ordinary; therefore the methods of analysis suitable to ordinary gear trains apply.
3. Take the algebraic sum of Steps 1 and 2 for each member of the train. The result is the number of revolutions made by that member for each +1 revolution of the arm. The Step 3 result, of course, should be +1 for the arm itself and 0 for the fixed gear.

The following simple example illustrates this procedure.

Example 8-5. If the arm of the train in Fig. 8-16 is rotating at 180 rpm counterclockwise, what is the angular velocity of the planet gear?

Solution. The following table has been filled in accordance with the three-step procedure.

Step	A	S	P
1	+1	+1	+1
2	0	−1	+2
3	+1	0	+3

In Step 1 the train was locked and given +1 turns. In Step 2 the arm was fixed, the sun gear released, and this gear given −1 turns. Since the planet is half the size of the sun and the mesh is external, the planet turns +2 revolutions for each −1 of the sun (regarding the train now as ordinary). The results in Step 3 show that, in the epicyclic operation of the train, the planet makes +3 turns for each +1 turn of the arm. Therefore the planet velocity corresponding to a velocity of 180 rpm ccw of the arm is 540 rpm ccw.

The method illustrated in the above problem is simply an application of Eq. 8-7. In any column of the table the Step 1 value is ω_A, the Step 2 value $\omega_{G/A}$, and the Step 3 value ω_G. For convenience in analyzing epicyclic trains with one fixed gear, the angular velocity of the arm is taken to be +1 arbitrarily so that the result of the tabular analysis will be the train value rather than the actual output speed.

Example 8-6. Determine the angular velocity of the arm in Fig. 8-17 if the driver is gear *S* and rotates at 880 rpm cw.

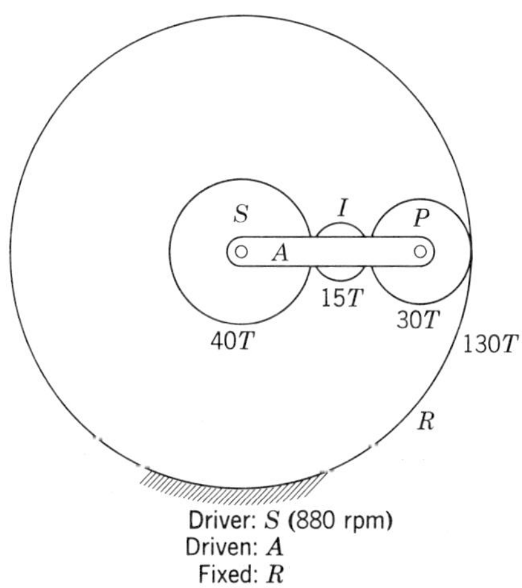

Fig. 8-17. Epicyclic train.

Solution. The tabular portion of the solution is:

Step	*A*	*R*	*P*	*I*	*S*
1	+1	+1	+1	+1	+1
2	0	−1	−13/3	+26/3	−13/4
3	+1	0	−10/3	+29/3	−9/4

The train value is $\omega_A/\omega_S = +1/(-\frac{9}{4}) = -\frac{4}{9}$. And therefore $\omega_A = (-\frac{4}{9})(-880) = +391$, that is, the arm rotates at 391 rpm ccw.

Example 8-7. For the train in Fig. 8-18, what is the angular velocity of gear S_2 if the arm is turning at 1000 rpm counterclockwise?

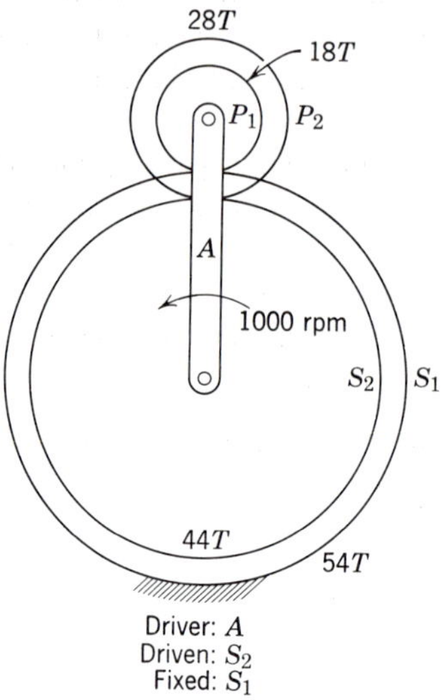

Fig. 8-18. Epicyclic train.

Solution

Step	S_1	P_1	P_2	S_2	A
1	+1	+1	+1	+1	+1
2	−1	+3	+3	−21/11	0
3	0	+4	+4	−10/11	+1

Therefore, $$\frac{\omega_{S_2}}{\omega_A} = \frac{-10/11}{+1} = -\frac{10}{11}$$

and $$\omega_{S_2} = -(10/11)(1000) = -909$$

Gear S_2 rotates at 909 rpm counterclockwise.

Trains with no fixed gears. For these trains the following procedure is used, based explicitly on Eq. 8-7. In general, the absolute angular velocities of two members of the train must be known.

1. In each column of the table enter the angular velocity of the arm.
2. The absolute velocity of some gear will be known. Employing Eq. 8-7, calculate the velocity of this gear relative to the arm. That is,

$$\omega_{G/A} = \omega_G - \omega_A$$

Fix the arm and rotate this gear at the calculated velocity. Treating the train as an ordinary train, calculate the corresponding velocities of the other gears.

3. Take the algebraic sum of Steps 1 and 2 for each member of the train (for each column in the table).

Example 8-8. For the epicyclic train in Fig. 8-19, determine the angular velocity of gear S if the arm turns at 50 rpm ccw and gear R at 10 rpm cw.

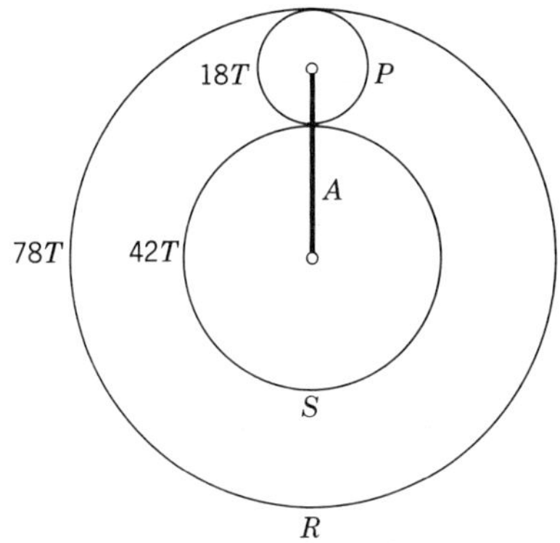

Fig. 8-19. Epicyclic train with no fixed gear.

Solution

Step	A	R	P	S
1	50	50	50	50
2	0	−60	−259	111
3	50	−10	−209	161

In Step 2 of the solution, it was reasoned that, since $\omega_R = -10$ rpm and $\omega_A = 50$,

$$\begin{aligned}\omega_{R/A} &= \omega_R - \omega_A \\ &= (-10) - (50) \\ &= -60\end{aligned}$$

As the table shows, the result is that gear S turns at 161 rpm counterclockwise.

Example 8-9. For the train of Fig. 8-19, what will be the arm velocity for a velocity of 75 rpm cw of S and a velocity of 125 rpm ccw of R?

Solution. For Step 1, since the angular velocity of the arm is unknown, the symbol "ω_A" is inserted. The absolute velocities of gears R and S are known. They have been inserted in Step 3. In Step 2 the arm is fixed, and the train is treated as an ordinary train. Therefore,

$$\frac{\omega_{S/A}}{\omega_{R/A}} = \frac{N_R}{N_S}$$

But
$$\omega_{S/A} = \omega_S - \omega_A = -75 - \omega_A$$

and
$$\omega_{R/S} = \omega_R - \omega_A = 125 - \omega_A$$

Therefore,
$$\frac{-75 - \omega_A}{125 - \omega_A} = \frac{78}{42}$$

or
$$\omega_A = 358$$

Step	A	R	P	S
1	ω_A	ω_A	ω_A	ω_A
2	0	$125 - \omega_A$		$-75 - \omega_A$
3	ω_A	125		-75

Example 8-10. For the train in Fig. 8-20, determine the angular velocity of shaft 2 if shaft 1 turns at 100 rpm ccw as viewed from the left.

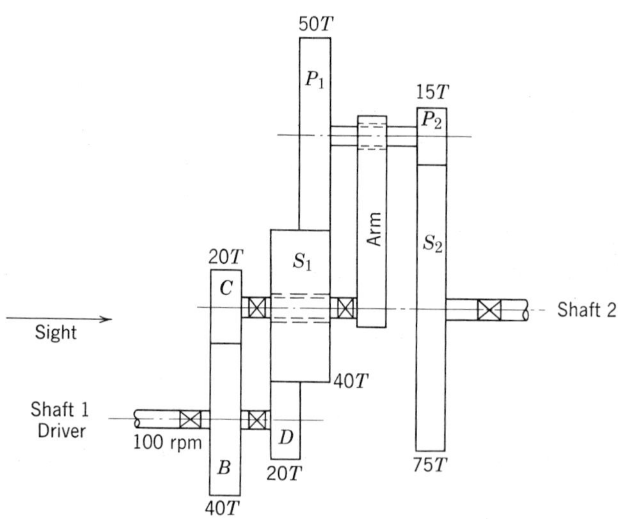

Fig. 8-20. Epicyclic train with no fixed gear.

Solution. In this train, gears B and D are not part of the epicyclic section; they merely drive it.

Now $\omega_{arm} = \omega_C = -40/20(100) = -200$

and $\omega_{S_1} = -20/40(100) = -50$

Therefore,
$$\begin{aligned}\omega_{S_1/A} &= \omega_{S_1} - \omega_A \\ &= (-50) - (-200) \\ &= 150\end{aligned}$$

Step	Arm	S_1	P_1, P_2	S_2
1	−200	−200	−200	−200
2	0	+150	−120	+24
3	−200	−50	−320	−176

Shaft 2 turns at 176 rpm clockwise as viewed from the left.

8-7. Gear Differentials

A *gear differential* is a device which adds or subtracts angular displacements. Since it is an epicyclic gear train with no fixed gears, it has two

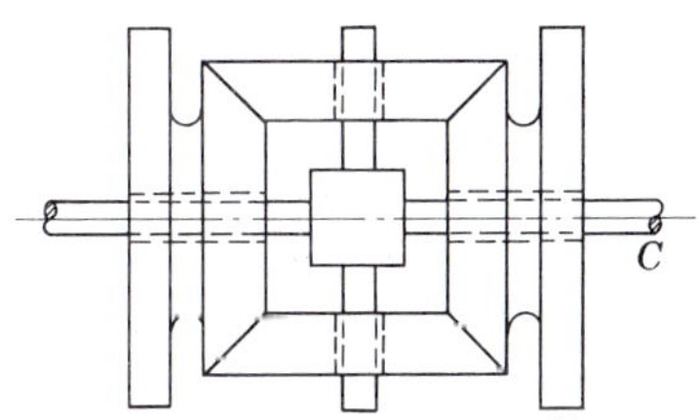

Fig. 8-21. Bevel gear differential.

degrees of freedom: two shaft inputs determine the output of the third shaft. An advantage of the gear differential as an addition device is its ability to add or subtract continuously, changing its output as the inputs change.

For the bevel gear differential of Fig. 8-21, controlling the motions of two of the three members, A, B, and C, ensures a definite and predictable motion for the third. The relationship between the three velocities can be found by applying the methods of the previous article, since the gear differential is an epicyclic gear train with no fixed gears. This has been done below, regarding A and C as the input gears and B as the output. In Step 2, when the spider shaft is regarded as fixed, gears A and B will turn in opposite directions, but at the same speed, since they are mounted on bevel gears which are of the same size.

Step	A	B	C
1	ω_C	ω_C	ω_C
2	$\omega_A - \omega_C$	$\omega_C - \omega_A$	0
3	ω_A	$2\omega_C - \omega_A$	ω_C

From the table it is evident that

$$\omega_B = 2\omega_C - \omega_A$$

or

$$\omega_A + \omega_B = 2\omega_C$$

This can be generalized as

$$a + b = 2c \tag{8-8}$$

where a, b, and c can represent either the angular velocities or the angular displacements of gears A, B, and C.

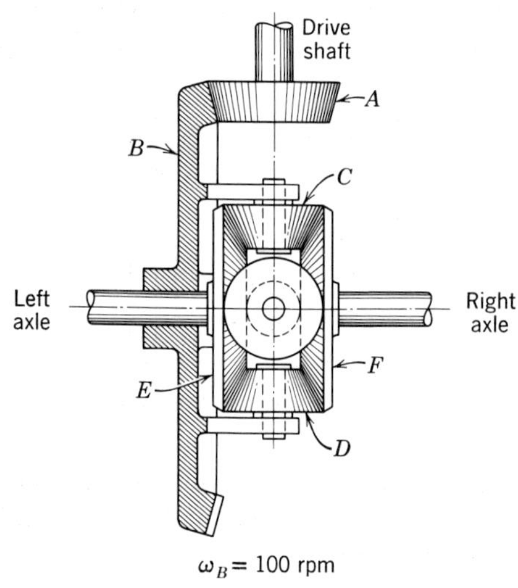

Fig. 8-22. Automobile differential.

A familiar application of the bevel gear differential is in the automobile, Fig. 8-22. The hypoid ring gear, B, is also the spider shaft or planet carrier for the differential. Hence, by Eq. 8-8, the sum of the axle speeds is always twice the ring gear speed. For example, when the car is moving on a straight road, if the ring gear speed is 600 rpm, the speed of each axle will be 600 rpm. But if the car then rounds a curve and the left axle speed drops to 500 rpm, the right axle speed will increase to 700 rpm if there is no change in the ring gear speed.

A gear differential can be used in solving an equation of the form

$$px + qy = rz$$

in control or computing applications. A gear train for accomplishing this is illustrated in Fig. 8-23. The number of turns made by shaft X represents

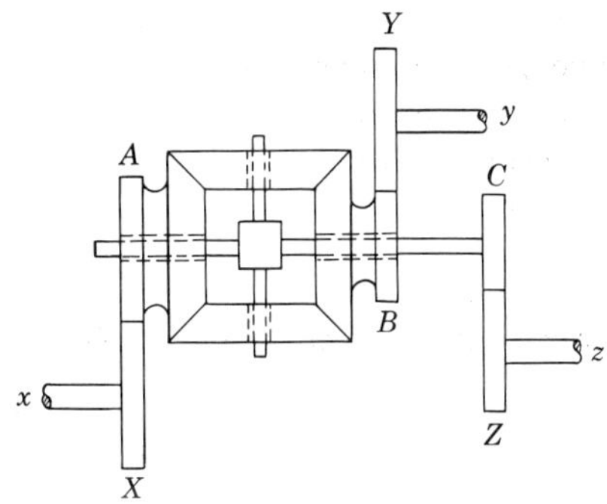

Fig. 8-23. Gear differential computing mechanism.

x; the turns made by Y represent y, and the turns by Z represent z. Let a be the number of turns made by A, b by B, and c by C. Then, Eq. 8-8 states that

$$a + b = 2c$$

But

$$\frac{a}{x} = \frac{X}{A}, \quad \frac{b}{y} = \frac{Y}{B}, \quad \text{and} \quad \frac{c}{z} = \frac{Z}{C}$$

where A, B, C, X, Y, Z represent the numbers of teeth or the pitch diameters of the gears mounted on the corresponding shafts. Solving the above equations for a, b, and c and substituting in Eq. 8-8, the result is

$$\frac{X}{A}x + \frac{Y}{B}y = 2\frac{Z}{C}z \tag{8-9}$$

Example 8-11. For the train in Fig. 8-23, if $X = 19$, $A = 43$, $B = 23$, $Y = 19$, $C = 17$, and $Z = 43$, what equation does the train solve?

Solution. Substituting the numbers of teeth in Eq. 8-9 results in

$$\frac{19}{43}x + \frac{19}{23}y = 2\frac{43}{17}z$$

$$0.442x + 0.826y = 5.06z$$

In this equation, the quantities x, y, and z can be either positive or negative since each of the three shafts can be turned in either a counterclockwise or a clockwise direction.

8-8. Designing Gear Trains

Thus far in this chapter, attention has been given primarily to those analyses of gear trains where the essential train dimensions were already known and the train values were sought. But a major task of the gear train designer is to determine the gear sizes necessary to achieve a given train value. And, unless systematic methods are adopted for train design, the designer will be forced to rely on tedious and time-consuming trial- and error-processes. The techniques to be explained below reduce the trial- and error-work to a minimum.

Factoring. Suppose an ordinary gear train is to be designed for an input speed of 1767 rpm and an output speed of 696 rpm. The train value is

$$\frac{\omega_o}{\omega_i} = \frac{696}{1767}$$

or, reducing to lowest terms,

$$\frac{\omega_o}{\omega_i} = \frac{232}{589}$$

Theoretically, at least, a two-gear pair could be used in which the driver has 232 teeth and the driven 589 teeth. But, in most applications, these gears would be impractically large. As an alternative, a four-gear compound train might be used, where A and C are the drivers and B and D the drivens. The train value is

$$\frac{\omega_o}{\omega_i} = \frac{232}{589} = \frac{N_A \times N_C}{N_B \times N_D}$$

It is necessary to determine two whole numbers, N_A and N_C, whose product is 232, as well as two other whole numbers, N_B and N_D, whose product is 589.

The initial step is to factor both numerator and denominator. Tables of factors and primes are very helpful (Refs. 1, 3, 7). The result is

$$\frac{232}{589} = \frac{2^3 \times 29}{19 \times 31} = \frac{8 \times 29}{19 \times 31}$$

Suppose further that each gear in the train shall have no fewer than 18 nor more than 100 teeth. Multiplying the numerator and denominator by 3, the result is

$$\frac{3 \times 8 \times 29}{3 \times 19 \times 31} = \frac{24 \times 29}{57 \times 31}$$

A four-gear compound gear train can be used if the drivers have 24 and 29 teeth while the driven gears have 57 and 31. So far as train value is concerned, it does not matter which driver and which driven are meshing nor which driver is first. For reasons of economy, the tooth loads should be minimized. Since the loads on teeth are inversely proportional to the pitch diameters, the first driver should be the largest of the drivers to be used and the last driven gear should be the largest of the drivens being used. The following rule can be used in mounting the gears of a compound ordinary gear train:

> Mount the driving gears in descending order of magnitude and the driven gears in ascending order.

For this example, then, the first driver should have 29 teeth and the last driven gear 57 teeth.

Example 8-12. It is required to design an ordinary gear train whose train value is 442/4123. Find suitable gear sizes. Each gear is to have no fewer than 18 nor more than 100 teeth.

Solution. A single gear pair cannot be used, for each gear would be too large. Perhaps a four-gear compound train will be more practical. Factoring the numerator and the denominator of the train value, the result is

$$\frac{442}{4123} = \frac{2 \times 13 \times 17}{7 \times 19 \times 31}$$

Since, if the denominator were regrouped into a product of two factors, one would be greater than 100, a four-gear train will not do. A six-gear train will have to be attempted. A quick inspection of the factored form of the train value reveals that a six-gear train is practical. However, each of the factors in the numerator is too small. To rectify this, the 2 will be multiplied by 9, the 13 by 2, and the 17 by 2. Then the denominator also will have to be multiplied by these same factors. The result so far is

$$\frac{18 \times 26 \times 34}{9 \times 2 \times 2 \times 7 \times 19 \times 31}$$

As a final step, the denominator must be regrouped into a product of three factors such that each is between 18 and 100 inclusive. One possible solution is

$$\frac{\omega_o}{\omega_i} = \frac{18 \times 26 \times 34}{63 \times 38 \times 62}$$

The drivers have 34, 26, and 18 teeth while the drivens have 38, 62, and 63 teeth.

Unfortunately, it is seldom so easy to find the appropriate values for numbers of teeth as it was in the above examples. Suppose the output speed of a gear is to be 266 rpm and the input speed 1561 rpm. Further, each gear is to have between 18 and 100 teeth inclusive. The train value is

$$\frac{\omega_o}{\omega_i} = \frac{266}{1561}$$

Factoring,

$$\frac{\omega_o}{\omega_i} = \frac{2 \times 7 \times 19}{7 \times 223}$$

It is useless to go on. One of the driven gears will have 223 teeth. An ordinary gear train having the exact desired train value, but with no gears having more than 100 teeth, is not possible.

In most cases some tolerance, however small, will have to be permitted on the train value. Then it will be possible to find any number of fractions whose values fall within the prescribed train value limits. Each of these fractions can be factored until one is found to be suitable.

A number of methods have been developed for finding a fraction approximating, within some prescribed tolerance, a required train value. Discussions of many of these methods will be found in Refs. 5, 8, 9, 10, 12. Brief discussions of the gear ratio logarithm and the conjugate fraction methods will be given now.

Gear ratio logarithms. The numerical value of the train value for an ordinary train can be found by setting the product of the numbers of teeth on the drivers over the product of the numbers of teeth on the drivens. For example, for a six-gear compound train in which *A*, *C*, and *E* are the drivers and *B*, *D*, and *F* the drivens, the numerical value of the train value is

$$TV = \frac{N_A N_C N_E}{N_B N_D N_F}$$

This can be rewritten as

$$TV = \left(\frac{N_A}{N_B}\right)\left(\frac{N_C}{N_D}\right)\left(\frac{N_E}{N_F}\right)$$

where it will be assumed that *B* is the gear meshing with *A*, *D* with *C*, and *F* with *E*. Taking the logarithms of both sides,

$$\text{Log } TV = \log \frac{N_A}{N_B} + \log \frac{N_C}{N_D} + \log \frac{N_E}{N_F}$$

This analysis applies, in general, to any ordinary train:

The logarithm of the train value (or speed ratio) for the train is the sum of the logarithms of the train value (or speed ratios) for the individual meshing pairs.

A *gear log table* lists the logarithms of gear ratios. For example, the gear log table in *Machinery's Handbook* lists in order of magnitude the various ratios greater than one that can be formed from the numbers 16 to 120. The first entry is 120/16 and the last is 120/119. The use of the gear log table is best shown by an example.

Example 8-13. Select gears for a four-gear train having a speed ratio 5.195 ± 0.002. Each gear shall have between 16 and 120 teeth. The speed reduction in the first meshing pair should be about twice that in the second pair.

Solution. Let A and C be the drivers and B and D the drivens. The numerical value of the speed ratio can be found by setting the product of the numbers of teeth on the driven gears over the product of the numbers of teeth on the drivers. Then the speed ratio is

$$SR = \frac{N_B \times N_D}{N_A \times N_C} = \left(\frac{N_B}{N_A}\right)\left(\frac{N_D}{N_C}\right)$$

Let

$$\frac{N_B}{N_A} = 2\left(\frac{N_D}{N_C}\right)$$

Then

$$5.195 = 2\left(\frac{N_D}{N_C}\right)^2$$

or

$$\left(\frac{N_D}{N_C}\right)^2 = 2.5975$$

Taking logarithms of both sides,

$$2 \log \frac{N_D}{N_C} = \log 2.5975$$

$$\log \frac{N_D}{N_C} = 0.20728$$

The closest value to this logarithm in the gear log table of *Machinery's Handbook* is

$$\log\left(\frac{108}{67}\right) = 0.20735$$

Tentatively, then, it will be assumed that $N_C = 67$ and $N_D = 108$. Since

$$\log SR = \log \frac{N_B}{N_A} + \log \frac{N_D}{N_C}$$

it follows that

$$\log \frac{N_B}{N_A} = \log SR - \log \frac{N_D}{N_C}$$
$$= \log 5.195 - \log \left(\frac{108}{67}\right)$$
$$= 0.50824$$

The nearest value to this in the table is the logarithm of the ratio 87/27. Therefore, try $N_B/N_A = 87/27$ and $N_D/N_C = 108/67$. The resulting speed ratio is

$$SR = \frac{87 \times 108}{27 \times 67}$$
$$= 5.194$$

which falls within the prescribed limits. If the final result had fallen outside the prescribed limits, a slightly different fraction for N_D/N_C could have been tried.

Mencke (Ref. 9) has pointed out that, when extremely small tolerances are prescribed in the selection of gears for a four-gear train, it is wise to select one ratio such that the second will fall within an area of the table where the gear ratio logarithms change value slowly. These latter areas are in the vicinity of the logarithms of the integral ratios 2/1, 3/1, 4/1, etc. If a first selection of a pair of ratios results in a train value or speed ratio just outside the prescribed limits, a subtle change can be made in it by selecting a new second ratio from among those in the close vicinity of the first choice.

Conjugate fractions. A *proper fraction* is one whose value is less than one, and whose numerator and denominator are integers containing no common integral factors other than one. For example, 13/17 and 14/25 are proper fractions, while 75/125 and 13/11 are not.

Two fractions, a/b and c/d, are said to be *conjugate* if

$$ad - bc = \pm 1 \tag{8-10}$$

Two conjugate fractions are the two of all proper fractions having the given denominators which are closest in value to each other. For example, 5/18 and 7/25 are conjugate. Of all possible pairs of proper fractions, such that one has 18 as its denominator and the other has 25 as its denominator, the pair 5/18 and 7/25 are closest in value to each other. Their difference

$$\frac{7}{25} - \frac{5}{18} = \frac{1}{450}$$

is the smallest possible difference between proper fractions having these denominators.

A *Brocot Table* is made up of all the proper fractions, with their decimal equivalents, whose denominators do not exceed some fixed number called the *base* of the system. They are listed in numerical order. For example, a Brocot table to base 120 has 1/120 as its first entry and 119/120 as its last. Conjugate fractions and the Brocot table have a number of interesting and useful properties which form the basis of a powerful technique for designing gear trains where extreme precision is necessary. The technique is explained in detail by Merritt (Ref. 10).

For the more usual applications, a technique based on two properties of conjugate fractions and the Brocot table can be used.

1. Of the infinite number of fractions lying between a pair of conjugate proper fractions, the one formed by setting the sum of the numerators of the original fractions over the sum of their denominators, and reducing to lowest terms, has the smallest numerator and the smallest denominator.
2. Any two consecutive fractions in a Brocot table are conjugate.

A Brocot table to base 120 can be found in Ref. 1, and one to base 60 in Ref. 7. A fragment of a 60-base table has been given in Table 8-3. The use to which the table can be put is illustrated in the following examples.

TABLE 8-3

PORTION OF A BROCOT TABLE TO BASE 60

Fraction	Decimal Equiv.	Fraction	Decimal Equiv.	Fraction	Decimal Equiv.
1/3	0.333333	5/14	0.357143	21/55	0.381818
20/59	0.338983	19/53	0.358490	13/34	0.382353
19/56	0.322034	14/39	0.358974	18/47	0.382979
18/53	0.339623	9/25	0.360000	23/60	0.383333
17/50	0.340000	13/36	0.361111	5/13	0.338462
16/47	0.340426	17/47	0.361702	22/57	0.385965
15/44	0.340909	21/58	0.362069	17/44	0.386364
14/41	0.341463	4/11	0.363636	12/31	0.387097
13/38	0.342105	19/52	0.365385	19/49	0.387755
12/35	0.342857	15/41	0.365854	7/18	0.388889
11/32	0.343750	11/30	0.366667	23/59	0.389830
10/29	0.344828	18/49	0.367347	16/41	0.390244
19/55	0.345454	7/19	0.368421	9/23	0.391304
9/26	0.346154	17/46	0.369565	20/51	0.392157
17/49	0.346939	10/27	0.370370	11/28	0.392857
8/23	0.347826	13/35	0.371429	13/33	0.393939
15/43	0.348837	16/43	0.372093	15/38	0.394737
7/20	0.350000	19/51	0.372549	17/43	0.395349
20/57	0.350877	22/59	0.372881	19/48	0.395833
13/37	0.351351	3/8	0.375000	21/53	0.396226
19/54	0.351852	20/53	0.377358	23/58	0.396552
6/17	0.352941	17/45	0.377778	2/5	0.400000
17/48	0.354167	14/37	0.378378		
11/31	0.354839	11/29	0.379310		
16/45	0.355556	19/50	0.380000		
21/59	0.355932	8/21	0.380952		

Example 8-14. Design a gear train for a train value of 2.69070 ± 0.00036. Gears having between 16 and 120 teeth are to be used.

Solution. Since a Brocot table contains only ratios with values less than one, the design will be based on the speed ratio.

$$SR_{\text{min}} = \frac{1}{2.69106} = 0.371601$$

$$SR_{\text{max}} = \frac{1}{2.69034} = 0.371700$$

By reference to Table 8-3, it is seen that the minimum and maximum values of the speed ratio both fall between the consecutive fractions

$$f_1 = \frac{13}{35} = 0.371429$$

$$f_2 = \frac{16}{43} = 0.372093$$

As a first trial, set the sum of the numerators over the sum of the denominators of these fractions. The result is

$$f_3 = \frac{13 + 16}{35 + 43} = \frac{29}{78} = 0.371795$$

This new fraction does not fall within the prescribed limits. The process of search will have to continue. The fractions now at hand are, arranged in order,

$$f_1 = \frac{13}{35} = 0.371429$$

$$f_3 = \frac{29}{78} = 0.371795$$

$$f_2 = \frac{16}{43} = 0.372093$$

The required fraction must lie between f_1 and f_3. The next trial is, therefore,

$$f_4 = \frac{13 + 29}{35 + 78} = \frac{42}{113} = 0.371681$$

This fraction is satisfactory. Since the speed ratio for the gear pair is 42/113, the driving gear must have 113 teeth and the driven gear 42 teeth.

Example 8-15. Design an ordinary gear train for a train value between 0.347390 and 0.347440 inclusive. Each gear should have at least 20 but no more than 90 teeth.

Solution. Consulting a Brocot table (see Table 8-3), the required train value limits fall between the fractions

$$f_1 = \frac{17}{49} = 0.346939$$

$$f_2 = \frac{8}{23} = 0.347826$$

As a first trial

$$f_3 = \frac{17 + 8}{49 + 23} = \frac{25}{72} = 0.347222$$

Fraction f_3 is not satisfactory. The required fraction must lie between fractions f_2 and f_3. The next trial is

$$f_4 = \frac{8 + 25}{23 + 72} = \frac{33}{95} = 0.347368$$

This too is unsatisfactory. Since the required fraction lies between f_2 and f_4, a further trial is

$$f_5 = \frac{8 + 33}{23 + 95} = \frac{41}{118} = 0.347458$$

This fraction falls very close to the train value limits, but still not within them. Since the limits fall between fractions f_4 and f_5, a further trial is

$$f_6 = \frac{33 + 41}{95 + 118} = \frac{74}{213} = 0.377417$$

This fraction falls within the limits and is satisfactory. A single gear pair will not be used; the driven gear would have to be too large (213 teeth). Factoring, the fraction can be written as

$$f_6 = \frac{74}{213} = \frac{2 \times 37}{3 \times 71} = \frac{20 \times 37}{30 \times 71}$$

A four-gear compound train can be used, where the drivers have 37 and 20 teeth and the driven gears have 30 and 71 teeth.

8-9. Flexible Connectors

Along with gears, belts and chains are the most frequently used devices to connect rotating shafts. Unlike gears, belts and chains can connect shafts relatively far apart with pulleys or sprockets whose diameters are independent of the center distance.

Belts are made flat, trapezoidal (V), ribbed, or toothed. They can operate at high efficiencies and at low levels of noise and vibration. And they need no lubrication. A flat-belt drive is shown in Fig. 8-24 and a V-belt drive in Fig. 8-25.

Chains can, in general, be used at higher loads than belts, but usually are restricted to connecting parallel shafts and require lubrication. Then,

Fig. 8-24. Flat-belt drive. (Courtesy Link-Belt Co., Chicago, Ill.)

too, they are likely to be noisier than belts. But chains share the advantage of gears and toothed belting in the ability to transmit load without slip while flat and V-belts are subject to some slip.

Flat belts. Open and crossed flat-belt drives are illustrated in Fig. 8-26. The speeds of the two pulleys in each case vary inversely as the effective pulley diameters. The effective pulley diameter is very nearly the sum of the pulley diameter and the belt thickness. Therefore the speed ratio for a flat-belt drive is

$$\frac{\omega_1}{\omega_2} = \frac{d_2 + t}{d_1 + t} \tag{8-11}$$

Fig. 8-25. V-belt drive. (Courtesy Link-Belt Co., Chicago, Ill.)

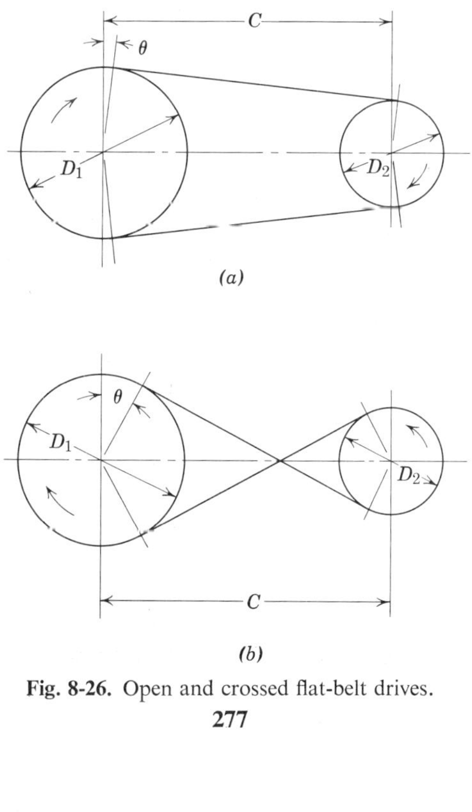

Fig. 8-26. Open and crossed flat-belt drives.

where pulley 1 will be regarded as the larger. The actual output speed will be somewhat less than that calculated by the above formula, due to some slip. The percentage slip is about 2 to 4% in most cases. The actual output speed is

$$\omega_{0\text{actual}} = \omega_{0\text{ideal}}\left(1 - \frac{s}{100}\right) \tag{8-12}$$

where s is the percent slip.

For the open belt drive of Fig. 8-26*a*, the belt length, L, is

$$L = 2\sqrt{C^2 - \left(\frac{D_1 - D_2}{2}\right)^2} + \frac{\pi}{2}(D_1 + D_2) + (D_1 - D_2)\arcsin\left(\frac{D_1 - D_2}{2C}\right) \tag{8-13}$$

where L = belt length, in.
C = center distance, in.
D_1 = diameter of larger pulley, in.
D_2 = diameter of smaller pulley, in.

For the crossed belt drive of Fig. 8-26*b*, the belt length is

$$L = 2\sqrt{C^2 - \left(\frac{D_1 + D_2}{2}\right)^2} + \frac{\pi}{2}(D_1 + D_2) + (D_1 + D_2)\arcsin\left(\frac{D_1 + D_2}{2C}\right) \tag{8-14}$$

where the nomenclature is the same as in the previous equation.

Example 8-16. An open flat-belt drive is to connect shafts 56 in. apart and rotating at speeds of 400 rpm (input) and 1253 rpm (output). The belt will be $\frac{3}{8}$ in. thick. Assuming a belt velocity of 4000 fpm and a slip of 4%, select stock pulley sizes for the drive, where the stock sizes are available in whole inch diameters. Calculate also the belt length.

Solution. The actual output speed will be less than the ideal output speed because of the slip. Since the slip is assumed to be 4%, the ideal output speed is, from Eq. 8-12,

$$\omega_{0\text{ideal}} = \frac{\omega_{0\text{actual}}}{1 - s/100} = \frac{1253}{1 - 4/100} = 1305 \text{ rpm}$$

Therefore pulley sizes should be selected in such a way that the theoretical output speed will be 1305 rpm.

Next the input pulley size will be estimated by assuming a belt speed of about 4000 fpm. Since

$$V = \frac{\pi\, d_1 n_1}{12}$$

the pulley diameter is

$$\begin{aligned} d_1 &= \frac{12V}{\pi n_1} \\ &= \frac{(12)(4000)}{(\pi)(400)} \\ &= 38.2 \text{ in.} \end{aligned}$$

The output pulley size can be found by solving Eq. 8-11 for d_2. The result is

$$\begin{aligned} d_2 &= \frac{\omega_1}{\omega_2}(d_1 + t) - t \\ &= \frac{400}{1305}(d_1 + 0.375) - 0.375 \\ &= 0.3065\, d_1 - 0.2601 \end{aligned}$$

This equation can be solved by trial and error, substituting whole inch diameters in the neighborhood of 38 for d_1, solving for d_2, and rounding off d_2 to a whole number. The best solution will be the one in which d_2 calculates to a value nearest a whole number. The results are

d_1	d_2
36	10.77
37	11.08
38	11.39
39	11.69
40	12.00
41	12.31

The best solution is $d_1 = 40$ and $d_2 = 12$. As a check, the ideal output speed should be calculated by Eq. 8-11.

$$\frac{\omega_2}{\omega_1} = \frac{d_1 + t}{d_2 + t}$$

$$\begin{aligned} \omega_2 &= (400)\,\frac{40.375}{12.375} \\ &= 1305 \text{ rpm} \end{aligned}$$

This is the correct theoretical speed for the output shaft. It will not always be possible to get so exact a result when using stock pulleys.

The belt length is found by using Eq. 8-13.

$$L = 2\sqrt{56^2 - \left(\frac{40 - 12}{2}\right)^2} + \frac{\pi}{2}(40 + 12)$$

$$+ (40 - 12) \text{ arc sin } \left(\frac{40 - 12}{2(56)}\right)$$

$$L = 197.1 \text{ in.}$$

V-belt drives. A V-belt drive is made up of grooved sheaves connected by one or more endless belts having a trapezoidal cross section. The belts usually consist of twisted rayon cords embedded in a core of rubber

Fig. 8-27. Link-belt P.I.V. variable speed drive. (Courtesy Link-Belt Co., Chicago, Ill.)

surrounded by a wear-resistant cover. The sheaves generally are gray iron or steel.

Both the sheave's groove and the belt have straight sides, but the bending of the belt around the sheave causes the belt to bulge at the sides, exerting a strong pressure against the sides of the groove.

The details of belt selection for a given application are covered in the manufacturers' catalogs. These should be consulted in selecting a V-belt drive.

Variable speed drives. Gear transmissions make possible a variation of output speed in a finite range, but only in steps. Often, however, there is need for a device which will permit the output speed to be set to any value within a finite range.

One example of such a device is the Link-belt P.I.V. variable speed drive shown in Fig. 8-27. These drives have been used in numerous applications where machines were operating under varying conditions, but where accurate control of the flow of materials through processes was demanded.

The chain connects two pairs of grooved conical wheels. The drive's ratio is varied by simultaneously moving one pair of wheels apart and the other closer together, thus varying the effective wheel diameters. Moveable packs of steel laminations permit the chain to lock itself in the grooves of the conical wheels, preventing slip.

REFERENCES

1. Earle Buckingham, *Manual of Gear Design, Section* 1, The Industrial Press, New York, 1935.
2. E. S. Cheaney, C. C. Paulus, and W. C. Raridan, "The Design and Application of Belts, Chains, and Gears," ASME Paper 59-MD-5.
3. *C.R.C. Standard Mathematical Tables*, 11th ed., Chemical Rubber Publishing Co., Cleveland, 1957.
4. Louis Dodge, "Gear Trains for Multiple Speeds," *Product Engineering*, February, 1948.
5. Robert G. Gerber, "Four-Gear Combinations," *Machine Design*, May 26, 1960.
6. James A. Hall, "Gear Trains for Changeable Speed Ratios," *Product Engineering*, February and March, 1934.
7. *Machinery's Handbook*, 15th ed., 1954, 16th ed., 1959, The Industrial Press, New York.
8. C. T. McComb and W. U. Matson, "Four Ways to Select Change Gears—and the Faster Fifth Way," *Product Engineering*, February 15, 1960.
9. George M. Mencke, "Gear Ratio Logarithms," *Machine Design*, February 1953.
10. H. E. Merritt, *Gear Trains*, Sir Isaac Pitman and Sons, London, 1947.
11. G. W. Michalec, "Gear Differentials, Part 1," *Machine Design*, October 1955.
12. Gerald Smithson, "Converting Decimal Fractions to Actual Gear Ratios," *Product Engineering*, August 1952.

PROBLEMS

8-1. For the gear train of Fig. P8-1, determine the velocity of gear *D*.

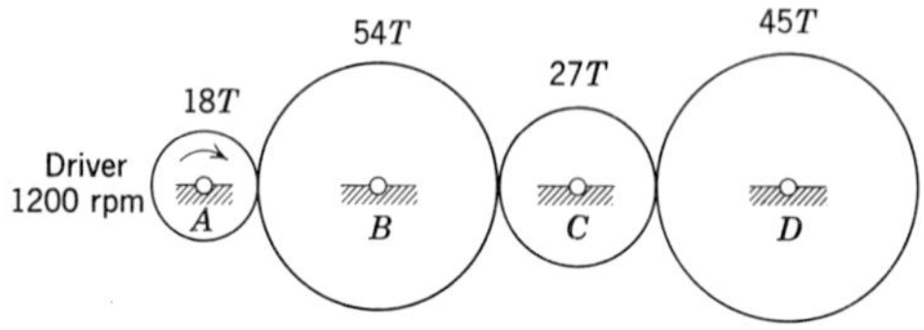

Fig. P8-1

8-2. For the gear train of Fig. P8-2, determine the velocity of gear *G*.

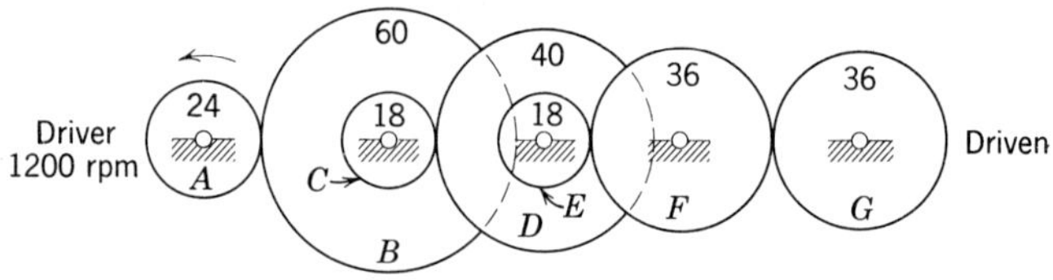

Fig. P8-2

8-3. What is the velocity of gear *F* in the train of Fig. P8-3?

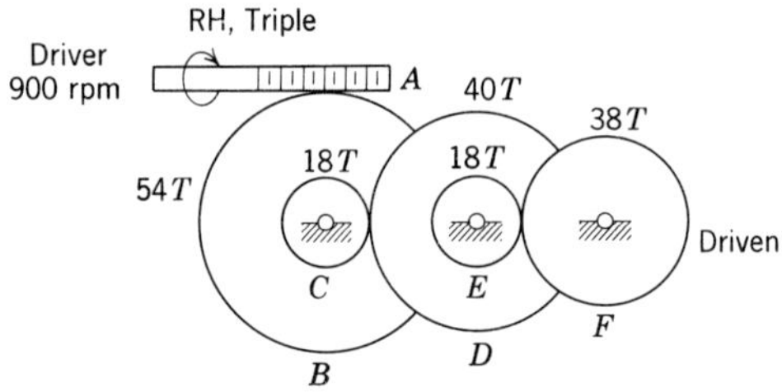

Fig. P8-3

8-4. A gear train operating a hoisting drum is shown in Fig. P8-4. What is the velocity of the weight at the instant shown?

8-5. A four-gear reverted train (Fig. 8-4) is to have a train value of 6. The gears of the first meshing pair are to have a diametral pitch of 6 while those of the second meshing pair are to be of 8 diametral pitch. Each gear should have at least 18 teeth. Select a suitable number of teeth for each gear.

8-6. The arm of the epicyclic gear unit of Fig. P8-6 is turning at 600 rpm clockwise. What is the velocity of gear *P*?

8-7. The arm of the epicyclic gear unit in Fig. P8-7 is turning at 600 rpm clockwise. What is the velocity of gear *S*?

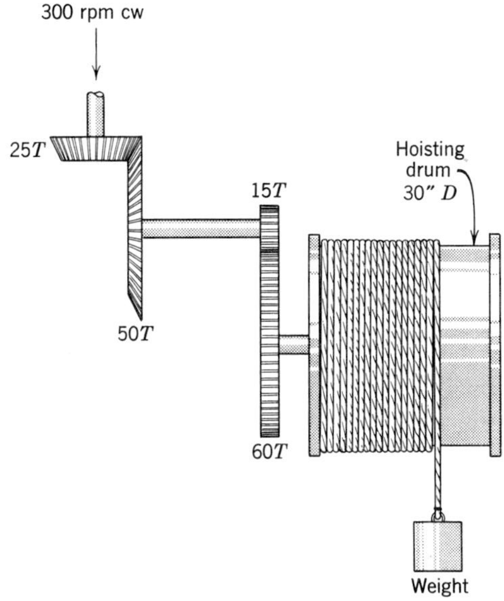

Fig. P8-4

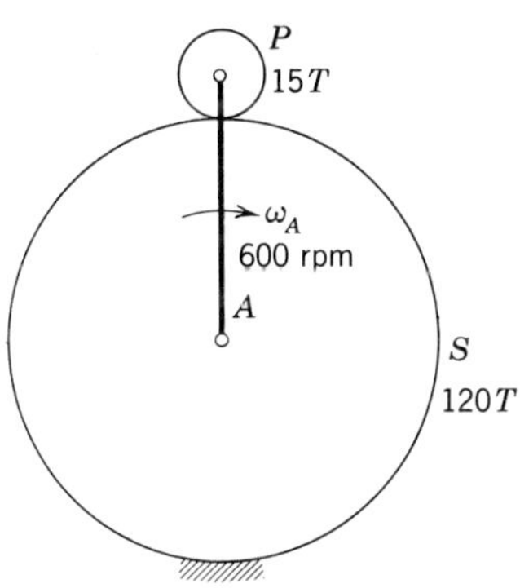

Fig. P8-6

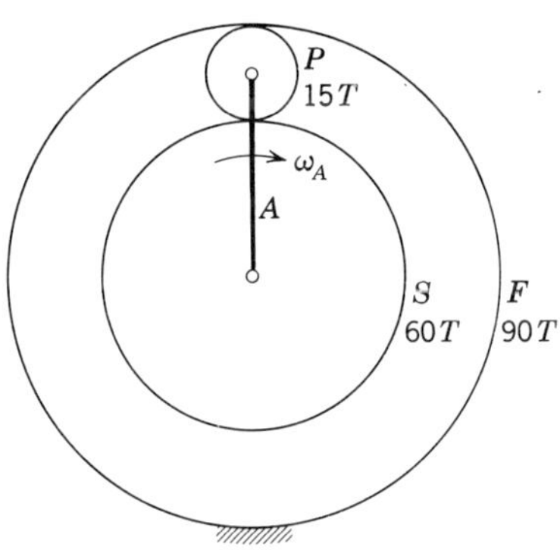

Fig. P8-7

8-8. What will be the angular velocity of the arm, A, in Fig. P8-8 if the gear E is fixed and B is turned at 200 rpm clockwise?

8-9. For the epicyclic train of Fig. P8-9 determine the number of teeth on gear B. What will be the velocity of E if B turns at 1800 rpm ccw and F is fixed?

8-10. What will be the velocity of gear D in Fig. P8-10 if A turns at 100 rpm ccw and B at 200 rpm ccw?

8-11. For the train of Fig. P8-11, determine the velocity of gear F if the arm turns at 250 rpm ccw and B at 50 rpm cw, viewed from the left.

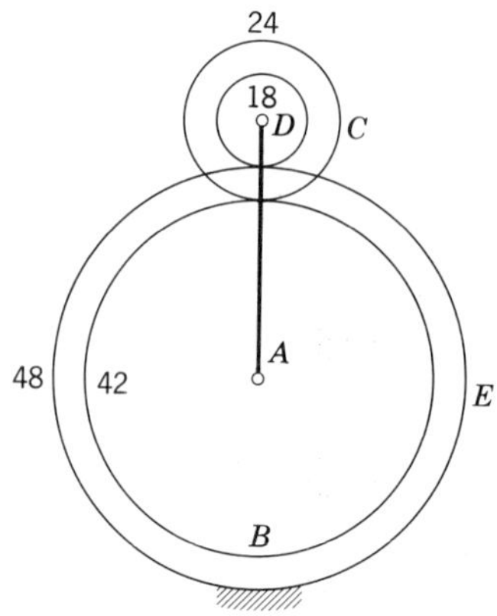

Fig. P8-8

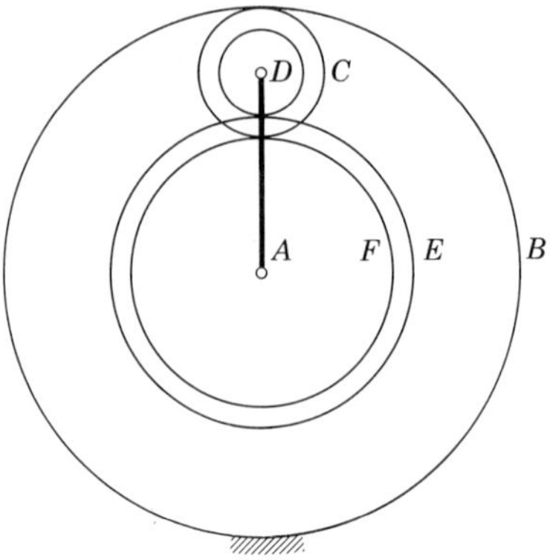

$N_C = 24$
$N_D = 18$
$N_E = 60$
$N_F = 54$

Fig. P8-9

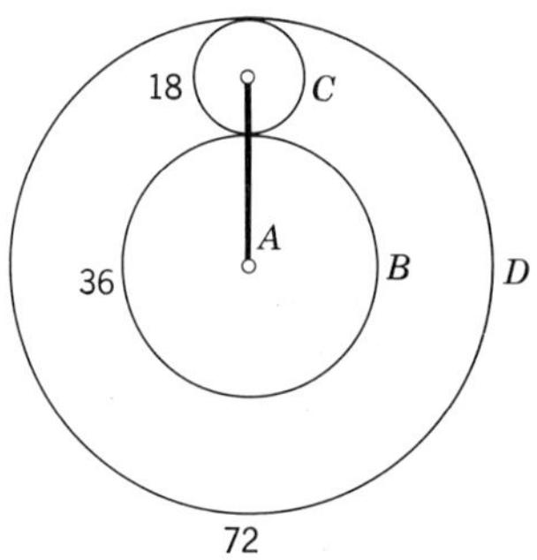

Fig. P8-10

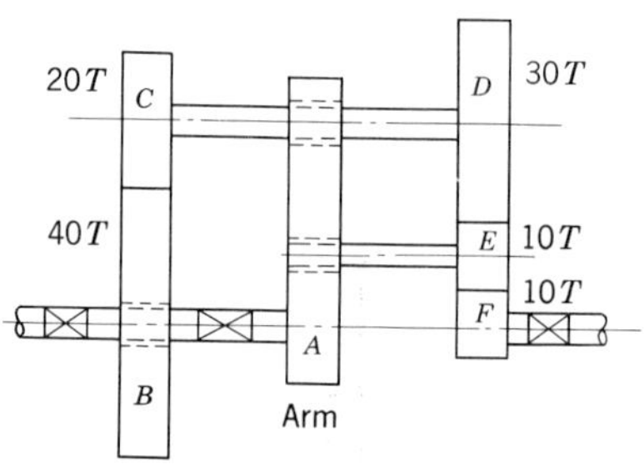

Fig. P8-11

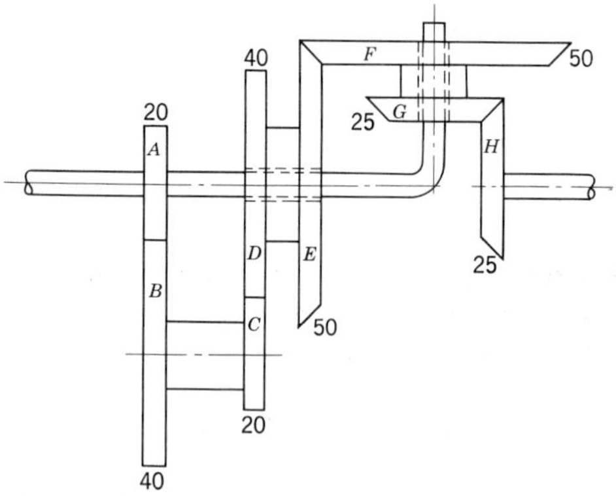

Fig. P8-12

8-12. Determine the velocity of gear H in Fig. P8-12 if A turns at 600 rpm cw, viewed from the left.

8-13. Select numbers of teeth for a four-gear train which is to have a train value of 323/341.

8-14. Design an ordinary train for a train value between 0.25716 and 0.25717.

8-15. Design an ordinary train for a train value of π with a permissible error of $\pm 0.01\%$.

8-16. A flat belt and pulley drive is to connect shafts 62 in. apart. The input speed is 600 rpm and the output speed is to be as close to 119 rpm as practical. Assume a slip of 3% and a belt velocity of about 4000 fpm. Select pulley diameters and calculate the belt length for a crossed drive. Stock pulleys are available in whole inch diameters.

Index